Also by Bruce Schechter

The Path of No Resistance:
The Story of the Revolution in Superconductivity

BRUCE SCHECHTER

A TOUCHSTONE BOOK
PUBLISHED BY SIMON & SCHUSTER

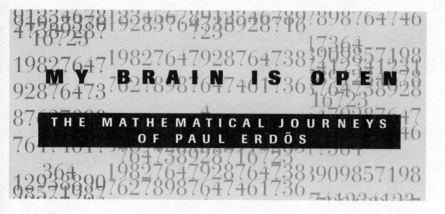

MY BRAIN IS OPEN

THE MATHEMATICAL JOURNEYS
OF PAUL ERDŐS

TOUCHSTONE
Rockefeller Center
1230 Avenue of the Americas
New York, NY 10020

Copyright © 1998 by Bruce Schechter
All rights reserved,
including the right of reproduction
in whole or in part in any form.
First Touchstone Edition 2000

TOUCHSTONE *and colophon are registered trademarks*
of Simon & Schuster, Inc.

Designed by Karolina Harris

Manufactured in the United States of America

1   3   5   7   9   10   8   6   4   2

The Library of Congress has cataloged the Simon & Schuster edition as follows:
Schechter, Bruce.
My brain is open: the mathematical journeys of Paul Erdős /
Bruce Schechter.
p.   cm.
Includes bibliographical references and index.
1. Erdős, Paul, 1913–   .
2. Mathematicians—Hungary—Biography.   I. Title.
QA29.E86S34   1998
510'.92—dc21

[b]   98-22293
                                                             CIP

ISBN 0-684-84635-7
0-684-85980-7 (Pbk)

Photograph of Babylonian clay tablet, page 32, is reprinted by permission of the
Columbia University Rare Book & Manuscript Library.
Photographs #1, 4, 6, 8, 14 & 15 reproduced with permission from the Center for
Excellence in Mathematical Education, 885 Red Mesa, Colorado Springs, CO 80906.

For my father,

Irving B Schechter

CONTENTS

ACKNOWLEDGMENTS

I learned about Paul Erdős from his very good friend Ronald Graham, whom I was profiling for *Discover* magazine, in 1982. Since then I always wanted to meet Erdős but never had the opportunity. When I learned of Erdős's death I remembered Graham's stories and resolved to find out more about this inspiring mathematician. I would like to give particular thanks to Ron, who is the repository for all things Erdős, for introducing me to so many members of the mathematics community.

Erdős's life touched many people, and I've relied on their memories, both written and spoken, to help me paint the portrait found in these pages. They include Yousef Alavi, Krishna Alladi, László Babai, Béla Bollobás, Neil Calkin, George Csisery, Martin Gardner, Janos Gellert, Dorian Goldfeld, Ronald Graham, Andrew Granville, Jerrold Grossman, George Helischauer, Melvin Henricksen, Shizuo Kakutani, Melvyn Nathanson, Janos Pach, Carl Pommerance, Bruce Rothschild, Richard Schelp, Allen Schwenk, John Selfridge, Miklós

Simonovits, Alexander Soifer, Vera Sós, Joel Spencer, Esther Szekeres, George Szekeres, Andrew Vazsonyi, and Laura Vazsonyi. I am indebted to Joel Spencer and Neil Calkin for their careful readings of the manuscript and thoughtful suggestions.

I am also indebted for able research to Nona Yates and Eszter Réti and for translations to Judi Sarossy. Robert Schechter, Susannah Greenberg, Jason Starr, and Catherine Keller provided valuable advice, support, and much-abused hospitality. I'm especially grateful to Bob Bender, my editor at Simon & Schuster, for his enthusiasm for this project and his always intelligent advice, and to my wonderful agent, Kris Dahl, for making it all happen.

Finally, my deepest gratitude and love go to my wife, Carmela Federico, for her brilliance, insight, editorial skills, and unwavering support.

CHAPTER ONE

TRAVELING

T H E call might come at midnight or an hour before dawn—mathematicians are oddly unable to handle the arithmetic of time zones. Typically, a thickly accented voice on the other end of the line would abruptly begin: "I am calling from Berlin. I want to speak to Erdős."

"He's not here yet."

"Where is he?"

"I don't know."

"Why don't you know?" *Click!*

Neither are mathematicians always observant of the social graces.

For more than sixty years mathematicians around the world have been roused from their abstract dreams by such calls, the first of the many disruptions that constituted a visit from Paul Erdős. The frequency of the calls would increase over the next several days and would culminate with a summons to the airport, where Erdős himself would appear, a short, frail man in a shapeless old suit, clutching two small suitcases that contained all of his worldly possessions. Stepping off the plane he would announce to the welcoming group of mathematicians, "My brain is open!"

Paul Erdős's brain, when open, was one of the wonders of the world, an Ali Baba's cave, glittering with mathematical treasures, gems of the most intricate cut and surpassing beauty. Unlike Ali Baba's cave, which was hidden behind a huge stone in a remote desert, Erdős and his brain were in perpetual motion. He moved between mathematical meetings, universities, and corporate think tanks, logging hundreds of thousands of miles. "Another roof, another proof," as he liked to say. "Want to meet Erdős?" mathematicians would ask. "Just stay here and wait. He'll show up." Along the way, in borrowed offices, guest bedrooms, and airplane cabins, Erdős wrote in excess of 1,500 papers, books, and articles, more than any other mathematician who ever lived. Among them are some of the great classics of the twentieth century, papers that opened up entire new fields and became the obsession and inspiration of generations of mathematicians.

The meaning of life, Erdős often said, was to prove and conjecture. Proof and conjecture are the tools with which mathematicians explore the Platonic universe of pure form, a universe that to many of them is as real as the universe in which they must reluctantly make their homes and livings, and far more beautiful. "If numbers aren't beautiful, I don't know what is," Erdős frequently remarked. And although, like all mathematicians, he was forced to make his home in the temporal world, he rejected worldly encumbrances. He had no place on earth he called home, nothing resembling a conventional year-round, nine-to-five job, and no family in the usual sense of the word. He arranged his life with only one purpose, to spend as many hours a day as possible engaged in the essential, life-affirming business of proof and conjecture.

For Erdős, the mathematics that consumed most of his waking hours was not a solitary pursuit but a social activity, a movable feast. One of the great mathematical discoveries of the twentieth century was the simple equation that two heads are better than one. Ever since Archimedes traced his circles in the sand, mathematicians, for the most part, have labored alone—that is, until some forgotten soul realized that mathematics could be done anywhere. Only paper and pencil were needed, and those were not strictly essential. A tablecloth would do in a pinch, or the mathematician could carry his equations in his head, like a chessmaster playing blindfolded. Strong coffee, and in Erdős's case even more powerful stimulants, helped too. Mathematicians began to frequent the coffeehouses of Budapest, Prague, and Paris, which led to the quip often attributed to Erdős: "A mathematician is a machine for turning coffee into theorems." Increasingly, mathematical papers became the work of two, three, or more collaborators. That radical transformation of how mathematics is created is the result of many factors, not the least of which was the infectious example set by Erdős.

Erdős had more collaborators than most people have aquaintances. He wrote papers with more than 450 collaborators—the exact number is still not known, since Erdős participated in the creation of new mathematics until the last day of his life, and his collaborators are expected to continue writing and publishing for years. The briefest encounter could lead to a publication—for scores of young mathematicians a publication that could become the cornerstone of their life's work. He would work with anyone who could keep up with him, the famous or the unknown. Having been a child prodigy himself, he was particularly interested in meeting and helping to develop the talents of young mathematicians. Many of the world's leading mathematicians owe their careers to an early meeting with Erdős.

Krishna Alladi, who is now a mathematician at the University of Florida, is one of the many young mathematicians whom Erdős helped. In 1974, when Alladi was an undergraduate in Madras, India, he began an independent investigation of a certain number theoretic function. His teachers could not help Alladi with his problem, nor could his father, who was a theoretical physicist and head of the Madras Institute of Mathematics. Alladi's father told some of his knowledgeable friends about his son's difficulty, and they suggested that he write to Erdős.

Because Erdős was constantly on the move, Alladi sent a letter to the Hungarian Academy of Sciences. In an astonishingly short time, Alladi heard from Erdős, who said he would soon be lecturing in

Calcutta. Could Alladi come there to meet him? Unfortunately, Alladi had examinations and could not attend, so he sent his father in his place to present the results of his research. After his father's talk, Alladi recounts, "Erdős walked up to him and told him in very polite terms that he was not interested in the father but in the son." Determined to meet with the promising young mathematician, Erdős, who was bound for Australia, rerouted his trip to stop briefly in Madras, which lies about 850 miles south of Calcutta.

Alladi was astonished that a great mathematician should change his plans to visit a student. He was nervous when he met Erdős at the airport, but that soon passed. "He talked to me as if he had known me since childhood," Alladi recalls. The first thing Erdős asked was, "Do you know my poem about Madras?" And then he recited:

> This is the city of Madras
> The home of the curry and the dhal,
> Where Iyers speak only to Iyengars
> And Iyengars speak only to God.

The Iyers and Iyengars are two Brahmin sects. The Iyers worship Shiva the Destroyer but will also worship in the temples of the Iyengars, who worship only Lord Vishnu, the Protector. Erdős explained that this was his variation on the poem about Boston and the pecking order among the Lowells, the Cabots, and God. Having put Alladi at ease, Erdős launched into a discussion of mathematics. Erdős was so impressed with Alladi, who was applying to graduate schools in the United States, that he wrote a letter on his behalf. Within a month Alladi received the Chancellor's Fellowship at the University of California, Los Angeles.

A CELEBRATED magazine article about Erdős was called "The Man Who Loved Only Numbers." While it is true that Erdős loved numbers, he loved much more. He loved to talk about history, politics, and almost any other subject. He loved to take long walks and to climb towers, no matter how dismal the prospective view; he loved to play ping-pong, chess, and Go; he loved to perform silly tricks to amuse children and to make sly jokes and thumb his nose at authority. But most of all, Erdős loved those who loved numbers, mathematicians. He showed that love by opening his pocket as well as his mind. Having no permanent job, Erdős also had little money, but whatever he had was at the service of others. If he heard of a graduate student who needed money to continue his studies, he would send a check. Whenever he lectured in Madras he would send his fee to the needy widow of the great Indian mathematician Srinivasa Ramanujan; he had never met Ramanujan or his wife, but the beauty of Ramanujan's equations had inspired Erdős as a young mathematician. In 1984 he won the prestigious Wolf prize, which came with a cash award of $50,000, easily the most money Erdős had ever received at one time. He gave $30,000 to endow a postdoctoral fellowship in the name of his parents at the Technion in Haifa, Israel, and used the remainder to help relatives, graduate students, and colleagues. "I kept only $720," Erdős recalled.

In the years before the Internet, there was Paul Erdős. He carried a shopping bag crammed with the latest papers, and his brain was stuffed with the latest gossip as well as an amazing database of the world of mathematics. He knew everybody: what they were interested in; what they had conjectured, proved, or were in the midst of proving; their phone numbers; the names and ages of their wives, children, pets; and much more. He could tell off the top of his head on which page in which obscure Russian journal a theorem similar to the one you were working on was proved in 1922. When he met a mathematician in Warsaw, say, he would immediately take up the conversation where they had left it two years earlier. During the iciest years of the Cold War Erdős's fame allowed him freely to cross the Iron Curtain, so that he became a vital link between the East and the West.

I N 1938, with Europe on the brink of war, Erdős fled to the United States and embarked on his mathematical journeys. This book is the story of those adventures. Because they took Erdős everywhere mathematics is done, this is also the story of the world of mathematics, a world virtually unknown to outsiders. Today perhaps the only mathematician most people can name is Theodore Kaczynski. The names of Karl Friedrich Gauss, Bernhard Riemann, Georg Cantor, and Leonhard Euler, who are to mathematics what Shakespeare is to literature and Mozart to music, are virtually unknown outside of the worlds of math and science.

For all the frequent-flier miles Erdős collected, his true voyages were journeys of the mind. Erdős carefully constructed his life to allow himself as much time as possible for those inward journeys, so a true biography of Erdős should spend almost as much time in the Platonic realm of mathematics as in the real world. For a layman this may seem to be a forbidding prospect. Fortunately, many of the ideas that fascinated Erdős can be easily grasped by anyone with a modest recollection of high school mathematics. The proofs and conjectures that made Erdős famous are, of course, far more difficult to follow, but that should not be of much concern to the reader. As Ralph Boas wrote, "Only professional mathematicians learn anything from proofs. Other people learn from explanations." Just as it is not necessary to understand how Glenn Gould fingers a difficult passage to be dazzled by his performance of the "Goldberg Variations," one does not have to understand the details of Erdős's elegant proofs to appreciate the beauty of his mathematics. And it is the nature of Erdős's work that while his proofs are difficult, the questions he asks can be quite easy to understand. Erdős often offered money for the solution to problems he proposed. Some of those problems are easy enough for readers of this book to understand—and perhaps even solve. Those who decide to try should be warned that, as Erdős has pointed out, when the number of hours it takes to solve one of his problems is taken into account, the cash prizes rarely exceed minimum wage. The true prize is to share in the joy that Erdős knew so well, joy in understanding a page of the eternal book of mathematics.

CHAPTER TWO

PROOF

OF ALL ESCAPES FROM REALITY, MATHEMATICS IS THE MOST SUCCESSFUL EVER. IT IS A FANTASY
THAT BECOMES ALL THE MORE ADDICTIVE BECAUSE IT WORKS BACK TO IMPROVE THE SAME
REALITY WE ARE TRYING TO EVADE. ALL OTHER ESCAPES—SEX, DRUGS, HOBBIES, WHATEVER—ARE
EPHEMERAL BY COMPARISON. THE MATHEMATICIAN'S FEELING OF TRIUMPH, AS HE FORCES THE
WORLD TO OBEY THE LAWS HIS IMAGINATION HAS CREATED, FEEDS ON ITS OWN SUCCESS. THE
WORLD IS PERMANENTLY CHANGED BY THE WORKINGS OF HIS MIND, AND THE CERTAINTY THAT
HIS CREATIONS WILL ENDURE RENEWS HIS CONFIDENCE AS NO OTHER PURSUIT.

—Gian-Carlo Rota, MIT mathematician

W H E N he was fourteen, during the years between the world
wars, Andrew Vazsonyi spent his days and nights in the back of his
father's shoe store in Budapest, ignoring all other needs and responsi-
bilities, filling page after page with diagrams and equations, solving
problems. He was hooked on mathematics. "I just absolutely became
addicted," he recalls. "That's the best word to use."

Mathematics addiction is reasonably common among the young.

Like music, mathematics exists in a separate world of form, relationship, and beauty, which is why history is full of mathematical and musical prodigies and is relatively devoid of toddling attorneys and infant stockbrokers. Children who are not old enough to cross the street can explore the infinite spaces of mathematics. But in Hungary during the early decades of this century, mathematical and scientific addiction had become an epidemic. The result was an astonishing generation of scientists and mathematicians that would change the world. The abstract musings of some of them, such as John von Neumann, Leo Szilard, and Edward Teller, would lead to the concrete achievements of atomic bombs and electronic computers. As a result their names would become famous even outside the world of academia. Less well known outside the arcane world of mathematics was a neighbor of Vazsonyi, a dark-eyed teenager named Paul Erdős, who would become one of the most brilliantly creative and inspirational mathematicians of all time.

Vazsonyi did not yet know Erdős, who was three years older, but he had heard about him. Budapest in the 1920s and 1930s was a bustling and cosmopolitan city of over a million, but for the tiny but influential Jewish community it was a small town. Vazsonyi's father visited Erdős's father and told him that his son was working day and night on math problems and would like to meet Paul. So one afternoon in 1930 there came a heavy knock on the shoe store door.

Kathy, the salesgirl, was astonished. Nobody ever knocked on a shopkeeper's door in the middle of the day! She opened the door, and in darted a small, energetic boy with a gait that reminded her of a gorilla. Kathy pointed him to the back room where Andrew was at work.

"Give me a four-digit number," Erdős demanded by way of introduction. Vazsonyi was startled, but this was an easy question. "2,532," Vazsonyi said.

"The square of it is 6,411,024. Sorry, I am getting old and decrepit and cannot tell you the cube," Erdős said. He was seventeen and had been calling himself old and decrepit for years.

"How many proofs of the Pythagorean theorem do you know?" Erdős next asked. Vazsonyi knew the celebrated theorem that related the sides of a right triangle; every schoolchild did. He knew a proof too, but only one. Who needed any more? "I know thirty-seven!" Erdős exclaimed.

The strange thing, Vazsonyi would recall more than sixty years later, is that Erdős was not boasting. "The concept was not applicable to him," Vazsonyi explained. Like the leader of an expedition who checks the equipment, supplies, and stamina of his companions before setting out to investigate rugged and uncharted continents, Erdős was merely determining what region of mathematical space he and Vazsonyi could best investigate together. After assuring himself of his companion's fitness for the journey, Erdős hastily outlined the proof of a famous and disturbing theorem of Georg Cantor that reveals, in a stroke, entirely unsuspected and unexplored realms, infinities within infinities. Having casually dropped that existential bombshell into the unsuspecting lap of Andrew Vazsonyi, Erdős said, "I must run," already halfway out the door, beginning another mathematical journey.

THE Danube River, a half-mile wide, stately and never blue, divides the city of Budapest into its two etymological halves, Buda and Pest. The royal palace is perched atop a cliff on the hilly and wooded Buda side, surrounding a giant equestrian statue of Prince Eugene of Savoy, who recaptured Buda from the Turks in 1686. This lofty vantage point affords an excellent view of the great bridges over the Danube to Pest, whose graceful catenaries soar like illustrations from a geometry book. At the turn of the century bankers, merchants, industrialists, artists, and intellectuals thronged the broad boulevards that ring Pest or rode beneath them in Europe's first subway. Between 1890 and 1900 the population of Budapest had increased by more than 40 percent to over three-quarters of a million souls, making it the sixth largest city in Europe. Because of Budapest's lively cafés, boulevards, parks, and financial exchange, visitors called it the "Little Paris on the Danube." What would not become apparent for years was that while the cafés were doing a booming business, the maternity wards of Budapest were churning out geniuses like a Ford assembly line.

Hungary's economic and intellectual flowering began with the Ausgleich of 1867, which established the dual monarchy with Austria. Under that agreement Hungary achieved something approaching independence from Austria; the Austrian Empire became the Austro-Hungarian Empire. With astonishing rapidity the engines of the industrial age and capitalism would transform Hungary. "The operators of those mechanisms," writes historian Richard Rhodes, "by virtue of their superior ambition and energy, but also by default, were Jews." Shortly after the establishment of the dual monarchy, discriminatory laws against Jews were repealed, opening to them all civic and political functions. The surge of Jewish immigration followed, paralleling the contemporaneous flood of Jewish immigrants from Russia to New York City.

Political power remained in the hands of the nobility, whose indifference to the gentile non-Hungarian minorities—nearly half the population—would keep a third of the gentiles illiterate as late as 1918, and most of them tied to the land. The Hungarian nobility, unwilling to dirty its hands on commerce, found allies in the Jews. By 1904 Hungarian Jews, who comprised about 5 percent of the population, accounted for about half of Hungary's lawyers and commercial businessmen, 60 percent of its doctors, and 80 percent of its financiers. Budapest Jews were also a dominant presence in the artistic, literary, musical, and scientific life of the country, which caused the growing anti-Semitic community to coin the derogatory label "Judapest."

The growing anti-Semitism would in later years cause many of the brightest members of the Hungarian Jewish community to flee their country. Some of the leading scientists and mathematicians, whose ideas and inventions would help form this century, were part of this tide of immigration. Among the better known were Leo Szilard, who was the first person to understand how chain reactions can unleash the power of the atom; John von Neumann, inventor of the electronic computer and game theory; and Edward Teller, the father of the hydrogen bomb. Less well known outside the world of science but equally influential were Theodor von Kármán, the father of supersonic flight; George de Hevesy, who received a Nobel Prize for his invention of the technique of using radioactive tracers that has had a revolutionary impact on virtually every field of science; and Eugene

Wigner, whose exploration of the foundations of quantum mechanics earned him a Nobel prize.

The list of great Hungarian scientists could be extended almost indefinitely, but even outside the sciences the prominence of Hungarians is extraordinary. In music it would include the conductors Georg Solti, George Szell, Fritz Reiner, Antal Dorati, and Eugene Ormandy, and the composers Béla Bartok and Zoltán Kodály. Hungarian visual arts in this century were dominated by László Moholy-Nagy, who founded the Chicago Institute of Design. Hollywood was even more influenced by the Magyar emigration. Movie moguls William Fox and Adolph Zukor were Budapest-born, as were Alexander Korda and his brothers, Vincent and Theodor, the director George Cukor, and the producer of *Casablanca*, Michael Curtisz. And of course Zsa Zsa Gabor and her sisters were Hungarian, as were Paul Lukas and Erich Weiss, better known as Harry Houdini.

Trying to account for what the physicist Otto Frisch called the "galaxy of brilliant Hungarian expatriates" is a favorite activity in scientific circles. The leading theory, attributed to the theoretical physicist Fritz Houtermans, is that "these people are really from Mars." Andrew Vazsonyi offers a particularly charming version of the extraterrestrial theory. "Well, at the beginning of the century," he says quite seriously, but with a twinkle in his eye, "some people from outer space landed on earth. They thought that the Hungarian women were the best-looking of all, and they took on the form of humans, and after a few years, they decided the Earth was not worth colonizing, so they left. Soon afterward this bunch of geniuses was born. That's the true story."

The actual explanation for Hungary's outpouring of genius is hard to find. Chance certainly played a role. But the strong intellectual values of the Jewish bourgeoisie, combined with the excellent Hungarian educational system, were the fertile field in which the random seeds of genetic chance could flourish.

Paul Erdős's family embodied the intellectual values and aspirations of Hungarian Jewry before World War I. His grandparents were observant Jews—probably, like most Hungarian Jews after 1867, "neologs," members of a movement of modernized Jews who observed Jewish traditions and holidays but rejected such orthodox practices as the shaving of women's heads. Erdős's father, the son of a schoolteacher in the town of Hódmezővárhely, was born Lajos Engländer on January 30, 1879. Although the Hapsburgs had formally agreed not to discriminate against Jews, they preferred not to be reminded of their presence; many Jews, therefore, prudently adopted Hungarian names. Lajos Engländer, a modest man, chose a modest name: Erdős, a common Hungarian name that means "from the woods." (Its approximate pronunciation is *air-dish*.)

Lajos Erdős was interested in mathematics and philosophy; like his father, he would become a high school teacher. He would, however, reject his father's religious observance as he had his name. He moved to Budapest to study mathematics at Pázmány University, now known as Eötvös University. There he made friends with Theodor von Kármán and with Lipót Fejer, who would become one of the greatest Hungarian mathematicians of his generation. He also met an attractive blue-eyed student from Vágbeszterce (now the town of Povazka Bystrica, Slovakia), Anna Wilhelm.

Anna Wilhelm, born on July 6, 1880, was the daughter of a religious shopkeeper. Her ambivalent observance of Jewish tradition ended one Yom Kippur with a visit from her fiancé. Yom Kippur is the holiest day of the Jewish year, a day devoted to fasting, repentance, and prayer. Lajos found Anna fasting but reading a volume of Maupassant. Like the good mathematician he was, he pointed out the apparent contradiction to Anna. After some reflection, Anna, a good mathematician herself, accepted this proof by *reductio ad absurdum*, burst into tears, and decided to give up her observance of Jewish traditions. It is not recorded whether, after drying her tears, she broke her fast.

For a Hungarian intellectual the decision to become a high school teacher was not the embrace of obscurity that it would have been in the United States. Hungarians have cherished their educational system since 996, when Adalbert the Saint, on his way from Prague to Rome, stopped in Hungary and founded the first Benedictine monastery and school in Hungary. For the next thousand years, throughout a troubled history, fine teachers, trained in Hungarian universities, helped preserve the cultural traditions of the nation.

Toward the close of the nineteenth century, József Eötvös, the Hungarian Minister of Culture, decided that, in order to face the challenges of the industrial age, he needed to introduce enlightened, secularized schools. To accomplish the task he hired Mór von Kármán, a leading educator and the father of Theodor von Kármán. Kármán built the Minta Gimnásium in Budapest, a model school along the lines of the German *Gymnasium* system. Kármán's model would become the basis of one of history's most successful educational systems. That accomplishment earned him a position in court, where he oversaw the education of the emperor's cousin. Franz Josef eventually awarded him a place in the hereditary nobility, bestowing upon him the predicate "von Szolloskislak," meaning "small grape," for the small vineyard near Budapest that Kármán owned. "I have shortened it to von," his son would write, "for even for me, a Hungarian, the full title is unpronounceable." The ennoblement of a schoolteacher is an indication of how highly the Hungarians valued education.

"According to a myth," George Marx writes, "all Martians [brilliant Hungarians] graduated from the same gymnasium as students of the same teacher." That is an exaggeration, but only a small one. The Nobel laureate chemists George de Hevesy and George Olah both graduated from the Roman Catholic school of the Piarist order, as did the physicist Roland Eötvös, whose work on the acceleration of falling bodies would help inspire Albert Einstein's theory of general relativity. Mór von Kármán's Minta Gimnásium produced, among others, his son, Theodor, and Edward Teller. John von Neumann and Eugene Wigner met in the mathematics class of László Rátz, a "miraculous" teacher at the Lutheran Gimnásium. "Our teachers were just enormously good," Wigner recalls, "but the mathematics teacher was fantastic. He gave private classes to Johnny von Neumann. He gave him private classes because he realized that this would be a great mathematician." In Budapest today there is a Rátz László Street (in Hungarian the last name is always written first), but none named for Wigner or von Neumann. Teaching high school, the career chosen by Erdős's parents, was honored in Hungary as in no other country. And the results showed.

Anna married Lajos in 1905, and almost immediately they began a family. They quickly had two girls, Magda and Klára, shining prodigies if a proud parent's memories can be trusted on such matters, the joy of the young couple's life. In March 1913 a happy Anna Erdős entered the hospital to give birth to the third addition to their growing family. But while she lay in the hospital an epidemic of scarlet fever tore through Budapest. On March 26 she gave birth to a son, Erdős Pál, later Westernized to Paul. By the time Anna brought Paul home from the hospital, her two daughters were dead. The devastated couple focused all of their love and energy on their dark-eyed baby. In later years Paul's mother would tell him that his sisters had been even smarter than he. If so, they must have been amazing, because by the age of four Paul was a full-blown mathematical prodigy.

Not all mathematicians start life as prodigies. Many, including some of the greatest, lead fairly ordinary childhoods and one day stumble over a fascinating problem, or book, or meet an inspirational teacher and become incurably hooked. Others, like Erdős, or the greatest mathematical prodigy of all time, Karl Friedrich Gauss, seem to have been born with hazy memories of the ideal Platonic realm of numbers that only needed refreshing. According to a historian of mathematics, Eric Temple Bell, Gauss was already known as a "wonder child" by the age of two; his "astounding intelligence impressed all who watched his phenomenal development as something not of this earth." When he was three, Gauss sat on a high stool watching his father add up a column of figures. When the elder Gauss wrote down the sum, Karl Friedrich, who had never been taught how to write numbers, add, or subtract, said, "Father, the reckoning is wrong, it should be . . ." A quick check proved that young Gauss was correct.

The most famous story of Gauss's unearthly precocity occurred when he was a ten-year-old student at a harsh school run by a Herr Büttner. To keep his young charges busy Büttner one day set them the task of adding up the numbers from 1 to 100. According to the legend, before Büttner finished stating the problem Gauss had flung down his slate and announced, *"Ligget se"*—there it lies. For the next hour, as the other children scratched away, adding, carrying, erasing and muttering, Gauss sat with his hands placidly folded. Büttner was ready to thrash the impudent child until he looked at the slate, which contained only one number. "To the end of his days Gauss loved to tell how the one number he had written was the correct answer and how all others were wrong," Bell writes. Gauss had instantly noticed that adding $1 + 2 + 3 + \ldots + 100$ is much easier if you consider the numbers out of order. He immediately understood that the first and last terms in this sequence, 1 and 100, add up to 101. The second and second to last terms, 2 and 99, also add up to 101. This pairing continues until 50 and 51, and no numbers are left over. In all there are 50 pairs of numbers that each add up to 101, so the sum of the numbers from 1 to 100 is $50 \times 101 = 5,050$. Gauss had discovered in a moment the trick for what mathematicians call "summing an arithmetic progression."

W H E N Erdős was three his mother, whom he would always call *Anyuka*, left him with the *Fräulein*, the hated German governess, while she was out of the house teaching. Paul learned to count in order to keep track of the days and months remaining until he could once again spend his days with his beloved *Anyuka* during her summer vacation. From counting it was a short step to arithmetic. Before very long he was impressing guests at the Erdős household by multiplying three- and four-digit numbers in his head.

In later years, Erdős would speak almost slightingly of his precocious calculating ability. Indeed, history is full of accounts of mental calculators whose feats would have left Paul in the dust but who in all other accomplishments were hopelessly backward. Jedediah Buxton, a well-known eighteenth-century calculating whiz, could square a thirty-nine-digit number in his head, though he rarely did: The task took him two and a half months. He amazed the clientele of his local pub with arithmetical legerdemain and was awarded with free pints of beer, of which he kept an exact tally.

Paul was not interested in bar tricks but in numbers and how they fitted together. The numbers that the three-year-old Paul could so easily manipulate are good for counting things like days, blocks, or pieces of cake. Mathematicians call these counting numbers the positive integers, the familiar sequence that begins 1, 2, 3 and never ends. Sumerian tablets dating back to the first half of the third millennium B.C. are incised with wedges that record the human fascination with the numbers that four thousand or so years later would be Paul's first toys.

Numbers are everyone's first toys. Cognitive scientists have recently discovered that infant brains come "prewired" with the ability to do simple arithmetic. "Mathematics," the MIT cognitive scientist Steven Pinker writes in his book *How the Mind Works*, "is part of our birthright." Evolution was the first mathematics teacher. Within the initial weeks of life babies already notice when the number of objects in a scene changes from two to three. By the time she is five months old a baby can do a simple sort of arithmetic. The psychologist Karen Wynn showed infants a Mickey Mouse doll and then hid it behind a screen. She then ostentatiously placed a second Mickey Mouse behind the screen. When Wynn lifted the screen the infants looked at the scene for a moment and lost interest; they had expected to see two dolls, and when they did they ceased to pay attention. Wynn then repeated the experiment, but this time she secretly removed one of the dolls before lifting the screen. When she revealed a solitary doll, the infants stared at the scene for a much longer time, amazed at Wynn's sleight-of-hand. To appreciate Wynn's trick the infants had to be able to understand that one doll plus one doll equals two dolls. In a variation of the experiment, Wynn hid two Mickeys behind the screen and then made a show of removing one. When she lifted the screen, the infants were amazed to see there were still two Mickeys when there should have been only one. This experiment implied that the infants had instinctively understood that two minus one equals one.

Psychologists have demonstrated that other mathematical concepts, like "greater than" and "less than," simple counting, arithmetic, and geometry are also innate in the very young. Hunter-gatherers did not need more advanced mathematics to survive, which is probably why the ability to solve partial differential equations is not hardwired in the human genome. This is sometimes a consolation to Erdős's close friend Ronald Graham, a mathematician and AT&T

Labs' Chief Scientist, when stumped by a tough mathematical problem. "Our brains were designed to keep us out of the rain, pick berries, and keep us from being killed. So the brain did that, but now it's got a whole new set of challenges—and we're getting better, but we're still a long way from being good at them."

"On evolutionary grounds it would be surprising if children were mentally equipped for school mathematics," Pinker writes, since this kind of math developed recently in a few modern cultures, too recently to have been encoded in our genome. School mathematics, a product of cultural evolution, required the development of such tools as language, reading, and writing to expand the abilities of the unaided human mind. Learning mathematics beyond the first intuitive steps is, therefore, largely a matter of hard work. "Without the esteem for hard-won mathematical skills that is common in other cultures," Pinker says, "the mastery is unlikely to blossom." Mathematical skill was held in higher esteem in Hungary than almost anywhere in the world, and it was held in higher esteem in the Erdős household than almost anywhere else in Hungary. When Paul started to show mathematical talent he was provided with plenty of incentive to blossom.

One day when Paul was four a visitor, after being amazed by Paul's instantaneous calculation of the number of seconds he had lived and other mathematical tricks, decided to try to throw the boy a curve. "What is 100 minus 250?" he asked.

Paul was quiet for a moment, briefly lost in a wonderful new space. Then he saw it, and delighted, shouted, "150 below zero!" It seems a small thing, but nobody had introduced Paul to negative numbers, a concept that cost mathematicians and philosophers thousands of years of acrimonious debate and strife. In an instant Paul had deduced that another sequence of numbers mirroring the counting numbers must exist. Still more impressive, he saw the delightful importance of this fact: His chest of mathematical toys, with the addition of negative numbers, had suddenly become infinitely larger. "It was an independent discovery," Erdős proudly explained.

An earlier age would have greeted Paul's discovery with incredulity. According to John Conway and Richard Guy, "When negative numbers were introduced, they were deemed impossible. What does it mean to speak of −3 apples? Of course −3 is not a 'real' number! But *now* it seems quite sensible to speak of negative temperatures and negative bank balances." The mathematician Leopold Kronecker once said, "God made the integers, all the rest is the work of man." The rest would become the life work of Paul Erdős; on that day he became a mathematician.

THE activities of mathematicians are obscure to outsiders. One frequent guess people hazard is that mathematicians spend their days thinking about numbers. Many, but by no means all, do. More generally—a phrase of which mathematicians are especially fond—mathematicians investigate the properties and relationships of "mathematical objects." Asking a mathematician to explain exactly what those are is a little like asking a poet what a poem is, or a musician what jazz is. Asked this last question, Louis Armstrong replied, "Man, if you gotta ask, you'll never know."

The passion and appreciation of the poet and the musician begin with nursery rhymes and melodies, and the passion for mathematics begins with counting. Numbers are the first and simplest mathematical objects.

The identities of the earliest mathematicians are lost in time, but from the Sumerian tablets it is clear that the impulse to create mathematics is ancient. Numbers were used to count cattle, measure fields, and construct calendars. At first people needed only whole numbers; no shepherd or insomniac counted fractional sheep. But fields, days, and drachmas could be divided, and fractions—ratios of whole numbers—bedeviled ancient schoolchildren. An important part of the history of mathematics has been expanding the concept of number. When the four-year-old Paul Erdős independently discovered the negative numbers, he was recapitulating some of the ancient history of mathematics.

Numbers may have been invented for calendars and commerce, but to the ancients they also revealed the patterns of the cosmos. In the hands of numerological mystics, numbers wove uncanny patterns that seemed to explain the order of the universe, though their explanations can seem tortured to modern minds. According to Plutarch, for example, the Egyptians held that the death of Osiris occurred on the seventeenth day of the lunar cycle, when the moon is waning. Pythagoras hated that number, not only for its association with the death of Osiris but because it was arithmetically flawed, which he proved by a bizarre bit of analysis. The problem with the seventeen, he explained, was that it lay "between the square number sixteen and the rectangular number eighteen, two numbers which alone of plane numbers have their perimeters equal to the areas enclosed by them." To Pythagoras all numbers had geometrical interpretations; plane numbers are the areas of rectangles and squares whose sides are whole numbers. A square measuring four units on a side has an area of $4 \times 4 = 16$, and a perimeter of $4 + 4 + 4 + 4 = 16$ units. A 6-by-3 rectangle has an area of 18 units, and a perimeter of $6 + 3 + 6 + 3 = 18$ units. As Pythagoras correctly observed, those are the only two plane figures with this property. A modern mathematician would consider things like that to be amusing algebraic exercises. Pythagoras felt such numerical patterns revealed the design of the cosmos. One historian remarked that because of his propensity toward this sort of numerological hocus-pocus Pythagoras was "one tenth . . . genius, nine-tenths sheer fudge." We can laugh at the fudge, but Pythagoras' genius gave birth to modern mathematics.

Pythagoras was born on the Greek island of Samos around the year 580 B.C. He became a wandering scholar, traveling extensively throughout Egypt, Babylon, and perhaps India. He probably visited most of the Seven Wonders of the ancient world and was exposed to most of its mysticism and cosmology. Upon his return he founded a secret brotherhood of mathematicians and mystics in southern Italy. The Pythagoreans led ascetic lives, ate no meat, and even eschewed beans because they resembled testicles. The Pythagoreans' battle cry was "number is everything." By number they meant the whole numbers and fractions made by taking ratios of whole numbers. Modern terminology enshrines this ancient hubris; these numbers are today known as rational numbers. The Pythagoreans worshiped their numbers and believed they were endowed with magical properties.

The oldest known document in number theory, Plimpton 322: a
Babylonian clay tablet incised between 1600 and 1900 B.C.E. The
tablet contains a table of so-called Pythagorean triplets, the lengths
of the sides of right triangles all of whose sides are whole numbers,
such as 3, 4, and 5 or 5, 12, and 13.

Indeed, most of this is just so much "fudge." But Pythagoras and his followers also worshiped rationality, and it is this rationality that led to their greatest achievement and to their ultimate humiliation.

Much of the mathematics known to Pythagoras had been kicking around the ancient world for a thousand years or more. Practical geometers had developed techniques to help measure fields, erect temples, and construct calendars, which worked beautifully. For example, Pythagoras' most celebrated claim to fame, the "Pythagorean theorem," beloved of high school geometry teachers and horribly mangled by the Scarecrow in *The Wizard of Oz*, was already known to the Babylonians.

In the 1920s archaeologists digging in the ancient Babylonian city of Senkereh uncovered a clay tablet incised with numbers and dating from between 1900 and 1600 B.C. At first they believed that it was the record of some commercial transaction, but in 1945 Otto Neugebauer and Abraham Sachs noticed that the numbers were actually

"Pythagorean triples," whole numbers that could be the sides of a right triangle. According to the Pythagorean theorem the sum of the squares of the legs of a right triangle is equal to the square of the hypotenuse. So if the legs of a right triangle happen to be 3 and 4 inches (the units are not important) the hypotenuse must be 5 inches, since $3^2 + 4^2 = 9 + 16 = 25 = 5^2$. The numbers 3, 4, and 5 constitute the smallest possible Pythagorean triple. It is easy to check that 5, 12, and 13 are another Pythagorean triple, and so are 7, 24, and 25. The tablet decoded by Neugebauer and Sachs, which is known as Plimpton 322, contains fifteen Pythagorean triples that, although the Babylonians had ten fingers like the rest of us, are oddly written in base sixty notation. Base sixty survives today in our method of dividing an hour into minutes and seconds and the division of a circle into 360 degrees. The tablet also contains a complete classification of all Pythagorean triples under certain conditions. The Babylonians clearly understood the Pythagorean theorem more than a thousand years before Pythagoras wrote it down. Why then is it not known as the Babylonian theorem? Did Pythagoras simply have a better press agent?

Pythagoras had something that the Babylonians did not; he had a proof. Before Pythagoras, assumptions and deductions were mixed together in a weird and unreliable brew. Observation and inspiration were the foundations of mathematics, but they were folded into the fudge in pre-Pythagorean days. Pythagoras changed all that by insisting that mathematics must proceed from a set of axioms—statements that seemed to him to be unarguably true—by the rules of logic to inevitable conclusions. The idea of proof was Pythagoras' most important contribution to mathematics; though successive generations of mathematicians have refined his notion, his fundamental insight remains the engine that drives mathematical progress. The Babylonians might have guessed at the relationship between the sides of a right triangle by measuring lots of triangles, adding up the squares of the lengths of the legs, and seeing that the sum always equaled the square of the hypotenuse. But no matter how many triangles they drew or how accurately they measured, they would never know if their law held exactly true for *every* possible triangle. Pythagoras was able to show, starting with assumptions nobody would think to question for thousands of years and using unarguable principles of logic, that the relationship observed by the Babylonians was exactly true for every right triangle in the universe. Using the deductive method sometimes attributed to his predecessor, the Greek geometer Thales of Miletus, Pythagoras ratified the Babylonians' observations by pure logic: he had *proved a theorem.* The teenage Paul Erdős would exult to Andrew Vazsonyi that he knew thirty-seven proofs of the Pythagorean theorem, each of them a thing of beauty. But one proof was enough; one proof changed everything.

Pythagoras would use Thales' precious brainchild, the method of logical proof, to examine the foundations of his numerical universe. The result would be Greek tragedy. The problem that devastated Pythagoras arose when he considered a simple right triangle with two legs, both exactly one unit long. How long is the hypotenuse of this triangle? Using the Pythagorean theorem nothing could be easier. The square of the hypotenuse is equal to the sum of the squares of the other two sides of the triangle. One squared plus 1 squared equals 2. Therefore, 2 is the square of the hypotenuse. That means the length of the hypotenuse is the square root of 2. According to Pythagoras this number, like all numbers, must be some fraction— the ratio of two whole numbers. Well, exactly what fraction is it?

To answer this question you can try doing what generations of mathematicians did before Pythagoras: guessing. A tablet in the Yale Babylonian collection, number 7289, records the remarkable guess that the square root of 2 is $1 + 24/60 + 51/60^2 + 10/60^3 = 1.41421296296296$. Nobody knows by what ingenious chain of logic the Babylonians arrived at this result, but a touch of a calculator key shows that it's extremely good—the actual square root of 2 is closer to 1.414213562373. Even that is not exact, just extremely close. To his horror and disgust, Pythagoras managed to prove that close is all you can ever get. No fraction, no matter how long you search or how clever you are, when multiplied by itself will equal exactly 2. With that proof, Pythagoras's numerological universe, which he preached was built entirely out of fractions—rational numbers—came tumbling down. Like Samson, it was his own strength, the strength bestowed upon him by his method of proof, that destroyed him.

Pythagoras, by an act of pure logic, was forced to admit that everything he believed about the world was wrong. The numbers he believed were everything could not even be used to measure the diagonal of a square! Erdős was delighted when he discovered that positive numbers weren't enough to describe his universe, but for him numbers were just toys. To Pythagoras and his followers numbers were literally everything, so they panicked. They sealed their tablets, they covered up. But the proof was too beautiful to be kept secret. Legend has it that a loose-lipped acolyte of the master divulged the proof and paid for his indiscretion with his life. According to Proclus Diadochus, a fifth-century scholar, "the guilty man, who fortuitously touched on and revealed this aspect of living things, was taken to the place where he began and there is forever beaten by the waves."

Pythagoras' beautiful proof is an example of what Erdős called a "book proof." "God has a transfinite book with all the theorems and their best proofs," Erdős was fond of saying. "I sometimes say, 'you don't really have to believe in God as long as you believe in the book.' Of course," he adds, "I do not really believe that The Book exists." Perhaps not, but Erdős's highest praise was to declare a proof to be "straight from The Book."

Whatever its existential status, the concept of a book of the best proofs lay at the heart of the philosophy that motivated his mathematical journeys. The best proofs, according to Erdős, are the simplest and most elegant, though he admits that "in some cases it's not really clearly defined." The great British mathematician G. H. Hardy stated: "Beauty is the first test: there is no permanent place in the world for ugly mathematics." Hardy was also begging the question; beauty is one of those vague, ill-defined words that mathematicians are supposed to hate. Logically, since all proofs are equally correct, one proof of a theorem is as good as another. As Gian-Carlo Rota, a mathematician and philosopher who has written about proofs and the concept of mathematical beauty, points out: "The expression 'correct proof' is redundant. Mathematical proof does not admit degrees. A sequence of steps in an argument is either a proof, or it is meaningless." Still, some proofs announce their correctness with an authority, clarity, inevitability, and simplicity that mathematicians sometimes call elegance. A great proof not only establishes the truth of the matter at hand but cuts to its heart. It enlightens.

In 1993 Andrew Wiles proved Fermat's Last Theorem, a problem that had baffled the best mathematicians—amateurs and professionals alike—for more than two centuries. Few people doubted Fermat's Last Theorem, which concerned solutions to an equation derived from generalizing the Pythagorean theorem, and almost nobody believed the theorem was of any real importance. Wiles's proof was hailed as a major achievement because not only had he finally pinned down the truth of Fermat's Last Theorem, but in doing so he had showed how branches of mathematics once thought to be unconnected are related and provided new techniques that will help mathematicians to solve problems about which they really do care.

Wiles's proof is beautiful, at least to the few mathematicians in the world who can understand it. But it is unlikely to be found in "The Book." Brevity in itself is not a necessary property of Book proofs; many Book proofs are long, but none are longer than they need to be. Insight, not length alone, is the true metric. Wiles's proof covers hundreds of pages and probably contains arguments that will be simplified as mathematicians understand their true spirit. "The first proof of a theorem can be excused for being clumsy," Erdős once said, having himself written clumsy proofs. Over time the essence of such proofs becomes clearer, and more concise and enlightening versions are written. Mark Twain understood this phenomenon when he apologized to a correspondent, "I didn't have time to write you a short letter so I wrote you a long one instead." Wiles's proof will undergo refinement by future generations of mathematicians, throwing off rays of enlightenment. It might even someday be cast in a form suitable for inclusion in The Book. Even so, it is unlikely that the proof will ever enlighten anyone but an expert.

Fortunately, a handful of proofs that belong in The Book can be understood by anyone with a vague memory of high school algebra and what Erdős would call a brain that is "open." Understanding such a proof is a little like viewing one of those three-dimensional pictures that appear at first glance to be nothing more than a sheet of marbled paper. You relax your eyes, open your mind, drop your prejudices, and concentrate. In a little while the surface of the paper seems to dissolve, revealing a three-dimensional image of a dolphin or a dinosaur. That moment is a revelation, magic, form emerging from formlessness. Doing mathematics can feel like that.

Although mathematicians each have their personal candidates for inclusion in The Book, they all agree that Pythagoras' proof that the square root of 2 is irrational is found in it, and probably on page one. The proof is swift and surprising, like a good joke or a magician's sleight-of-hand, yet it communicates something of the flavor of what it is to do mathematics. That is probably what motivated Erdős when one day in the 1970s he decided to explain Pythagoras' proof to Laura, Andrew Vazsonyi's wife.

Laura is a musician who studied the mathematics required to graduate from high school and no more. So she was surprised when Erdős, who had been staying with them on a visit, said: "Laura, I would like to explain to you the Pythagorean scandal."

"Okay, Erdős," she said a little warily. They often talked about many things—history, politics, and the state of Erdős's laundry—but never mathematics.

Erdős took out a blank piece of paper and said, "Laura, if you do not understand a step, let me know and I will clarify the proof, okay?" Laura nodded, and Erdős proceeded very slowly, in his thickly accented English.

Pythagoras' proof starts with a peculiarly daring gambit often employed by mathematicians: He assumed that what he was trying to prove was wrong. "It is a far finer gambit than any chess gambit," Hardy explains, because, "a chess player may offer the sacrifice of a pawn or even a piece, but the mathematician offers the *game.*" Pythagoras started by assuming the square root of 2 *was* a rational number. That is, he assumed that the square root of 2 is a fraction.

The next step is to capture this idea in symbols. What does it mean to say that something is a fraction? That's easy: A fraction is just the ratio of two whole numbers, like 17/12 or 577/408. The English sentence "The square root of 2 is rational" can therefore be symbolically written:

$$a/b = \sqrt{2}.$$

The letters a and b stand for any two whole numbers. Erdős explained to Laura that Pythagoras also required that this fraction be written in lowest terms. As everyone learns in grade school, the same fraction can be written in an infinite number of ways. For example, 17/12 is the same as 34/24 (top and bottom multiplied by 2) or 51/36 (top and bottom multiplied by 3). When a fraction is written in lowest terms its numerator and denominator—in this case, a and b—have no factors in common.

Since the fraction a/b is assumed to be equal to the square root of 2, this fraction squared is equal to 2; after all, that's what a square root is. So:

$$a^2/b^2 = 2.$$

Multiplying each side by b^2, this can be rearranged to read:

$$a^2 = 2b^2.$$

This equation simply states the obvious fact that if a fraction is equal to 2, the numerator is twice the denominator. But it says something more: The numerator, a^2, must be an *even* number, because it is 2 times b^2. The value of b^2 doesn't matter; 2 times any number is always an even number. If a^2 is even, then a must be even as well; if a were odd, then a^2 would also be odd, since an odd number times an odd number is always odd. "Okay, Laura?" Erdős asked, and she nodded. What does it mean to say that a is even? An even number is a number that is divisible by 2, so that any even number can be written as twice a smaller number. If a is even it can be written as twice some other number that we will call c. In symbols the assertion that a is even can therefore be written:

$$a = 2c.$$

We are really interested in a^2, not a, but that's not a problem. Just square the above equation and you find that:

$$a^2 = 4c^2.$$

In other words, any even number squared must be a multiple of 4. But we already showed that a^2 was equal to 2 times b^2. Combining these facts we can conclude that:

$$2b^2 = 4c^2.$$

Dividing both sides of this equation by 2 gives:

$$b^2 = 2c^2.$$

We are almost done. This equation says that b^2 must be an even number, since it is 2 times some other number. Using the same reasoning we used before, we conclude that if b^2 is even then b must also be even. At this point it would be surprising if Erdős did not shout "Aha!" or words to that effect, because the proof is done. We have proved that if a/b is equal to the square root of 2, both a and b must be even. But a and b *can't* both be even, since we started by insisting the fraction a/b be in lowest terms. A fraction with both an even numerator and an even denominator is never in lowest terms, since the numerator and denominator can both be divided by 2. This is a contradiction, which means that our initial assumption must be wrong.

"See! The assumption is wrong, the square root of 2 cannot be rational," Erdős concluded triumphantly.

His triumph was fleeting, however, because Laura did not like the proof. She felt as if she had been tricked. Erdős became angry and said, "I asked you to tell me at every step if you don't understand something. You said nothing!"

"Why didn't you tell me at the beginning that this is all wrong?" Laura replied, and Erdős stormed away. Vazsonyi was so amused by this failed lesson that he decided to keep the page with Erdős's explanation as a souvenir. "I recalled that when Albert Einstein gave one of his last talks, at the end they unscrewed the blackboard and sent it to the Smithsonian. So I asked Erdős to certify the document, so I could keep it for history."

It seemed to Laura that Erdős had been dishonest by not divulging that his initial statement was wrong. In fact, he had employed one of the most powerful and disconcerting weapons in the mathematician's toolbox, *reductio ad absurdum*. By reducing his initial assumption to an absurdity he had proved its opposite. Pythagoras had probably started more honestly than Erdős had, by assuming what he believed with all his soul was a correct statement: The square root of 2 could be written as some fraction. When he worked out the consequences of the statement, as Erdős had for Laura, he found the same contradiction. In other fields of human endeavor people try to sweep such contradictions under the rug, but mathematical logic leaves no place to hide. Pythagoras knew with awful certainty that his carefully constructed, rational universe was a fiction. Although he tried to keep it secret he could not deny the awful truth. Pure logic forced him to accept what his heart denied, that there were worlds beyond the world of his rational imagination.

Paul's discovery of negative numbers at age four affected him almost as deeply. But later that year he would make what he called "my second great discovery," which would haunt him for the rest of his life. While shopping with his mother it occurred to him that the sequence of years that constituted his life would not go on forever. Though time was infinite, his life was not. Later in his life he would become famous for so-called existence proofs, but this was a nonexistence proof. "I started to cry. I knew I would die," he said, "From then on I've always wanted to be younger."

Erdős would brood and joke about his own mortality for the rest of his life. He was particularly amused when scientists revised upward their estimation of the age of the earth. He once gave a lecture called "My First Two-and-a-Half-Billion Years in Mathematics." His justification? "When I was a child the earth was said to be two billion years old. Now scientists say it's four and a half billion. So that makes me two and a half billion. The students at the lecture drew a time line that showed me riding a dinosaur. I was asked, 'How were the dinosaurs?' Later, the right answer occurred to me: 'You know, I don't remember, because an old man only remembers the very early years, and the dinosaurs were born yesterday, only a hundred million years ago.' "

A few years before his sixtieth birthday, in the early 1970s, Erdős had started appending the initials PGOM to his name, which he explained stood for Poor Great Old Man. When he turned sixty he lengthened it to PGOMLD for Poor Great Old Man Living Dead; at sixty-five he added the letters AD for Archaeological Discovery. At seventy he tacked on LD for Legally Dead; and at seventy-five he added CD for Counts Dead. The last set was added because the Hungarian Academy of Sciences, in order to keep its membership roll fixed at two hundred, drops members at age seventy-five, though they can keep full privileges. When he was nearing seventy-five, Erdős explained to a reporter that those final initials would probably never be used. "Maybe I won't have to face that emergency," he explained. "[They] may have to be for my memory. I'm miserably old. I'm really not well. I don't understand what's happening to my body. Maybe the final solution." He was still one of the most prolific mathematicians in the world, traveling more than even the most jaded jet-setter, though he often ended a session of mathematical collaboration by saying, "We'll continue tomorrow . . . if I live."

The four-year-old Paul who first faced his own mortality was a handsome child, with dark eyes and a serious expression. His young parents, having suffered the tragic loss of their two daughters to scarlet fever, did everything they could to cosset and protect their precocious son. Disease was the least of their worries, for war was soon going to break out. Within a year of Paul's birth the Archduke Ferdinand was assassinated in Sarajevo and Austria-Hungary declared war on Serbia. Russia was next infected, declaring war on Austria-Hungary. War spread like fire to include Germany, France, and Britain. While Paul was discovering for himself the reality of death, men were everywhere being slaughtered in the Great War.

Within weeks of the outbreak of the war, tens of thousands of Hungarian men, including Lajos Erdős, were drafted into the military and sent off to battle the Russians on the Eastern front. Large numbers died on the battlefield. The wounded and captured were taken by the Russians on long forced marches to their internment. Lajos was one of them. He spent six hard years as a prisoner of war in Siberia.

The war marked the end of Hungary's golden age. Maimed soldiers filled the streets, and the poor flocked to soup kitchens established by the great Budapest industrialists. Throughout those years Anna Erdős was employed and was able to provide a comfortable life for Paul. Then Austria-Hungary finally lost the war in 1918, and the dual monarchy was dissolved. A new, independent Hungary under the leadership of Count Mihály Károly found no support from the West in its efforts to rebuild its economy and maintain its independence. Within a year Hungary was invaded on all sides by its neighbors Romania, Czechoslovakia, and Yugoslavia; a dispirited Károly resigned. He turned Hungary over to a small group of Hungarian Communists led by Béla Kún, who looked optimistically to the Soviet Union for support. Károly had looked to the West for help, while Kún looked East; both were chasing phantoms.

The Hungarian Commune was a disaster that lasted exactly 132 days. Its brief reign, the historian John Lukacs writes, was "marked by imbecility, inefficiency and terror." The Commune wasted no time in enacting its ideals—nationalizing the economy, separating church and state, and secularizing the schools. Opposition was crushed by a wave of violence known as the "Red Terror."

When the Hungarian Commune collapsed on August 2, 1919, a new and far more virulent terror, known as the "White Terror," swept the streets of Budapest. Under the new government of Hungary," headed by a former admiral and commander of the Austro-Hungarian fleet, Miklós Horthy Nagybánya, former Communists were beaten and hanged. Along with this increasing anti-Communist fervor came a rising tide of anti-Semitism. Kún and most of his commissars had been Jewish. By extension, all Hungarian Jews were viewed as having been associated with the former regime, although most had not been.

Erdős recalled standing with his mother on the fifth-floor balcony of their apartment, watching Jews being beaten on the streets below. In those days many Jews converted to avoid persecution. Anna asked her six-year-old son whether he thought they too should convert. Erdős and his family, like many Hungarian Jews of that age, were not observant. Erdős would later remark that he "never cared about being Jewish." Still, guided by the principled obstinacy that would someday put him at odds with governments on both sides of the Iron Curtain, he said: "You may do as you wish but I will remain the way I was born."

That a boy of six should express such a wish was admirable and remarkable. Even more remarkable was that a woman of forty should have unquestioningly acceded to his wishes. Anna's devotion to her son was almost legendary among their Budapest friends. She took care of all of his daily needs, anticipated all his desires, and protected him from all threats, real and perceived. She could have been the real-life model of the mother in the joke who insisted that her son be carried everywhere. One day, as they checked into a hotel, a woman watched as an attendant carried the boy across the lobby. "Oh, poor thing! Can't he walk?" she asked.

"Of course he can," the mother replied. "But thank God he doesn't have to."

Paul did not learn how to tie his own shoelaces until he was eleven years old. Ten years later, when he left home for the first time to study in England, he discovered that he had never actually learned to butter bread for himself. He hadn't had to. "I remember clearly," he said, "it was tea time and bread was served. I was too embarrassed to admit I had never buttered it. I tried. It wasn't so hard."

Anna Erdős was luckier than most during the days of the Commune, or so it seemed at the time. She was promoted to school principal. Under the Horthy regime an honor by the Commune was deemed a crime, and Anna was forever barred from teaching in public schools. She continued to make a living by tutoring private students. Then, in November 1920, Lajos Erdős, Paul's beloved *Apuka* (Hungarian for Daddy), finally returned from the war. The wear of enduring six years of Siberian cold, malnutrition, and civil war was written on his face. On first seeing him, Paul exclaimed, "*Apuka*, you are *really* old!"

Lajos Erdős resumed teaching mathematics and science at one of Budapest's top high schools, the Szent Isván Gimnásium. Between his salary and the money Anna earned tutoring and working as a technical editor, they maintained a comfortable middle-class existence. Erdős would recall that he had difficulty adjusting to the discipline of school. "I never liked, and still do not like tight restrictions," he would say. That may have been just the natural reaction of an overindulged child or the result of what today might be called hyperactivity—friends of his youth would recall him as excitable and agitated, arms constantly aflap, given to hopping out of chairs with no warning, racing across rooms and stopping suddenly, inches before impact with the wall. Still, when combined with his parents' fear of his catching a disease, it meant that Paul would remain at home to be educated by *Apuka* and *Anyuka*, "free of all the inconveniences that were a burden on other children." Only when he was older would Paul spend an occasional year or two at the Tavaszmezö gymnasium and at Szent Isván, where, as at home, he was his father's prize pupil.

But it would be another teacher, through the pages of the magazine he created, who would introduce Paul to the community of mathematics and mathematicians, the proofs and conjectures, men and women, who would become his family, his love, and his life.

CHAPTER THREE

CONTACT

"MATHEMATICS IS THE ONLY UNIVERSAL LANGUAGE THERE IS, SENATOR!"

— line spoken by Jodie Foster in the film Contact

W H E N *Apuka* returned to Budapest at last after his years in a Siberian prison camp, Paul's home schooling began in earnest. Math, of course, was the centerpiece of the curriculum, but Paul's parents, suspecting that his future might lie outside Hungary, placed an almost equal emphasis on foreign languages. Even though both parents spoke fluent German, they hired the hated *Fräulein* to care for Paul and to help him master that language. While a prisoner in Siberia Paul's father had distracted himself from the cold and hunger by learning to speak French and English. Upon his return he taught those languages to his son. Hungarians have a notoriously difficult time losing their accents when they speak foreign languages, even when schooled by native speakers. Lajos Erdős had to puzzle out

English pronunciations for himself, with the aid of the few cryptic diacritical hints in his primer. He passed those guesses on to his son, who added his own idiosyncrasies. Erdős would tell a Hungarian interviewer: "I was not even ten when I already spoke fluent English." Fluent, perhaps, and charming certainly, but so heavily accented that a filmmaker who made an English-language documentary about Erdős felt it necessary to use subtitles whenever Erdős was speaking. Erdős's friends were often amused when they overheard him giving English lessons to his mother. She might ask, "Palkó, how do you call the fruit 'szilva' in English?"

"Plimm, Mother, plimm," came his confident reply.

Erdős's parents also began to teach him the universal language of mathematics, which he rapidly learned to speak without a trace of an accent. Music and mathematics have both been nominated as universal languages, but mathematics has the better claim. In his own private jargon Erdős called music "noise," although he enjoyed listening to it. Music of one culture is often, if not exactly noise, hard to appreciate by another. At best music communicates emotions, but even then it is imprecise. While some might hear fate knocking on the door in the opening bars of Beethoven's Fifth Symphony, others could hear a demon laughing: "Ha, ha, ha, *haaa!*" A visitor from another planet, even if it could hear sound, would almost certainly not head for the nearest record store. In the event of a close encounter, attempting to sing our greetings would be futile. Mathematics would probably work better.

Mathematics, as Jodie Foster's character told the senator, is a universal language that we could justifiably expect extraterrestrials to speak. Pythagoras may have erred in insisting that all was number, but it is likely that numbers are understood by all. One plus one equaling two is as true on Betelgeuse as it is in Budapest. Regardless of how many fingers our extraterrestrial friends have, they will count in the same way.

Different cultures have different styles of doing mathematics; Hungarian mathematics has a flavor and style different from French or German. Which is a far cry from claiming, as a recent university course description did, that "mathematics is a social construct." In that view, known as ethnomathematics, Western mathematics, like music and literature, is as dependent on issues of gender, power, and politics as it is on pure reason. The implication, as Edward Rothstein pointed out in the *New York Times*, is "that the standard Western approach to mathematics is limited and incomplete, one approach among many."

Mathematicians are finite, flawed beings who spend their lives trying to understand the infinite and perfect. That kind of thing is bound to result in problems and misunderstandings. Trends and fashion, politics and pig-headedness all affect the lurching progress of mathematical knowledge. None of them, however, affect the validity of mathematical knowledge. "There are many ways," Rothstein writes, "to show that the ratio between the circumference of any circle and its diameter is always the same, a number known as pi. The priests, farmers, and builders who first used that ratio may have had various intentions and goals. And the ratio may be given names like pi or zed or Milwaukee, for that matter. But the number and its meaning are unaffected by cultural apparatus and influence."

So, if mathematics is a universal language, how do you say hello? Scientists concerned with SETI, the Search for Extraterrestrial Intelligence, have given this matter a lot of thought and decided that the best thing to do would be to treat our alien friends as if they were bright children, and begin to count at them. Testing, one, two, three.

Just counting—*beep, beep-beep, beep-beep-beep, beep-beep-beep-beep*—would probably work, though it is just barely plausible that some natural phenomena could produce such a sequence. SETI researchers, such as Carl Sagan, were convinced that the simplest message that "can be recognized as emanating unambiguously from intelligent beings" would consist of the first dozen or so prime numbers. It's hard to imagine a naturally occurring physical phenomenon that could produce such a sequence, and it's hard to imagine an intelligent species that would not understand it.

Humans have been fascinated by primes for as long as they could count. Primes are whole numbers that can be divided only by 1 and themselves. The first few primes are 2, 3, 5, 7, and 11 (by convention, 1 is not considered prime). The number 15 is composite—not prime

The First 100 Primes									
2	3	5	7	11	13	17	19	23	29
31	37	41	43	47	53	59	61	67	71
73	79	83	89	97	101	103	107	109	113
127	131	137	139	149	151	157	163	167	173
179	181	191	193	197	199	211	223	227	229
233	239	241	251	257	263	269	271	277	281
283	293	307	311	313	317	331	337	347	349
353	359	367	373	379	383	389	397	401	409
419	421	431	433	439	443	449	457	461	463
467	479	487	491	499	503	509	521	523	541

—since it is the product of 3 and 5. Primes are often called the atoms of arithmetic, since every number is either a prime or can be written as a product of smaller primes in exactly one way. (This, by the way, is not an observation but a theorem. It is true for all integers, but mathematicians have invented exotic species of numbers for which the theorem is not true.)

The arc of mathematical careers often echoes the history of mathematics. Primes were discovered when mathematics was young, and young mathematicians, like Paul Erdős, often start out by being fascinated by prime numbers. Determining whether a number is prime or composite is the kind of game that appeals to young calculating whizzes like Erdős. Eventually, maybe after factoring hundreds or thousands of numbers, the game is bound to wear thin, and the budding mathematician is likely to wonder if there is a largest prime or do the primes go on forever? Look at a table of prime numbers. Do you see a pattern? Don't feel bad if you don't; nobody sees a pattern. The first few dozen primes are found in irregular clumps spaced pretty close together. But as you move farther out, the average distance between primes seems to increase steadily. Sometimes you will find two or three large primes near each other, but the trend is clear: Primes become rarer the farther out you look. Are the primes drying up? Is there a largest prime number?

The largest number currently known to humans to be prime was discovered on January 27, 1998, by a nineteen-year-old University of

Southern California sophomore, Roland Clarkson. Clarkson found his 909,526-digit prime on an old Pentium 200-megahertz computer running a program written by George Woltman, a computer programmer from Florida. Woltman's program looks for what are known as Mersenne primes, named for Marin Mersenne, a French monk who studied them in the seventeenth century. Mersenne primes are primes that are one less than a power of 2, or, in symbols, $2^p - 1$. The number 3 is a Mersenne prime, since it is equal to $2^2 - 1$; so is 7 ($2^3 - 1$), 31 ($2^5 - 1$) and 127 ($2^7 - 1$). Until the year 1536 people thought that $2^p - 1$ was prime as long as p itself was a prime, but then Hudalricus Regius noticed that $2^{11} - 1 = 2,047$ was equal to 23×89.

By the twentieth century mathematicians had invented fast computer algorithms to check whether a Mersenne number is prime. Still, until recently searching for gigantic Mersenne primes was usually done by engineers who wanted to take their latest supercomputer on a shakedown cruise. In 1996 Woltman overturned the supercomputer's primacy when he launched the Great Internet Mersenne Prime Search (GIMPS). Thousands of computer enthusiasts around the world downloaded a copy of Woltman's program and reserved a block of Mersenne numbers to check for primality. The combined power of all those puny desktop machines, many of them outdated models saved from the scrapheap by Woltman's GIMPS project, exceeded the fastest supercomputers. Before long the GIMPS project had scored its first hit, a prime 420,921 digits long. Less than a year later the GIMPS had another success, and more computer enthusiasts around the world began to volunteer their spare computer time. Thanks to a clever program written by Scott Kurowski, a San Jose software development manager, the search for Mersenne primes became totally automated. Anyone with a PC can download the program, which runs quietly in the background, soaking up idle moments of computer time that would ordinarily go to waste. Roland Clarkson, who memorizes the digits of pi as a hobby, was the lucky lottery winner from among 4,000 participants whose computer earned him a place in the math history books. Clarkson knows that he will probably not be there for long. The search for the biggest prime will never end for the simple reason that there is no biggest prime. The number of primes is infinite.

· · ·

T H E proof that there is an infinity of primes is found in the pages of one of the most influential books on mathematics—or any other subject—ever written, *The Elements* by Euclid, the Greek mathematician who lived in the third century B.C. Before Euclid, mathematics was a jumble of seemingly self-evident statements and the deductions that could be made from them by the application of logic. Much progress had been made, and useful results had abounded, but the *ad hoc* nature of this method obscured the underlying relations among different results and made progress difficult.

Euclid essentially cleaned house. He started by defining basic objects, such as points, lines, planes, and so on. He then wrote down a set of statements concerning relationships among those objects that he viewed as so blatantly obvious that they required no proof. One such axiom is, for example, if two objects are both equal to the same object they are equal to each other. It seems both obvious and irreducible to simpler truths. If primes are the atoms of arithmetic, axioms are the atoms of reason. Euclid showed how, starting from these axioms and using simple logic, many of the geometrical and arithmetical truths known to the Greeks could be proved.

For two thousand years a knowledge of Euclid was considered to be an essential part of a good education. Thomas Hobbes somehow escaped exposure to *The Elements* until he was forty, but when he finally read it, he was astonished. It happened by accident, John Aubrey writes. One day, "being in a gentleman's library, Euclid's *Elements* lay open, and 'twas the *47 El. libri I* [Pythagoras' Theorem]. He read the proposition. 'By God,' sayd he, 'this is impossible': So he reads the demonstration of it, which referred him back to such a proposition; which proposition he read. That referred him back to another, which he also read. *Et sic deinceps,* that at last he was demonstratively convinced of that trueth. This made him in love with geometry." David Herbert Donald recounts that Abraham Lincoln, "believing, as did most of his contemporaries, that mental faculties, like muscles, could be strengthened by rigorous exercise, . . . secured a copy of Euclid's principles of geometry and with determination set himself to working out the theorems and problems. With quiet pride he reported in 1860 that he had 'studied and nearly mastered the six books of Euclid.' "

Bertrand Russell recalled his first exposure to Euclid at age eleven as "one of the great events of my life, as dazzling as first love. I had not imagined there was anything so delicious in the world. From that moment until I was thirty-eight, mathematics was my chief interest and my chief source of happiness."

Paul Erdős was a year younger than Russell when his father showed him Euclid's proof of the infinitude of primes; the bedazzlement would last him for the rest of his life. The proof, one of the most beautiful in all of mathematics, is definitely one for The Book. Euclid's strategy is similar to that which Pythagoras used to prove that the square root of 2 is irrational. He starts by assuming the opposite of what he wants to prove and sees where this will lead him. In other words, Euclid assumes that there *is* a greatest prime, which we will call P_N. (If the use of unknown quantities like P_N gives you the willies, *pretending* that the largest prime is 7 or 11 or some other small prime might help make the logic more transparent.) If this assumption leads to a contradiction, then the logical conclusion is that the assumption was incorrect: There is no largest prime.

The first step of the proof is to multiply all of the primes together to give one large number:

$$A = 2 \times 3 \times 5 \times 7 \times 11 \times \ldots \times P_N.$$

A is obviously divisible by every prime; that's just how we built it. Now let's add 1 to A and look at the resulting number, which we will call P.

$$P = A + 1 = (2 \times 3 \times 5 \times 7 \times 11 \times \ldots \times P_N) + 1.$$

P is either a prime or it isn't; those are the only choices. If P is a prime we are done, since P is clearly bigger than P_N, which contradicts the assumption that P_N is the largest prime.

Every number is either a prime or the product of primes. So if P is not prime it *must* be divisible by some prime, but which one?

Dividing P (which is A + 1) by 2, 3, 5, 7, or any of the other primes up to P_N clearly leaves a remainder of 1, since A, being the product of all other primes, must be divisible by each of them. If P is not prime, it must be divisible by *a prime that is greater than P_N*. But we assumed that there was no such number. So the assumption that P is not a prime also leads to a contradiction. Therefore, there can be no greatest prime; the number of primes is infinite.

If you took the trouble to read the above proof (and I hope you did) you probably found it slow going. Part of Euclid's legacy is that mathematical arguments tend to be terse and tightly wrapped, and almost always take some time to tease apart. Nobody can speed-read mathematics; to even the most fluent speakers it remains a foreign language. Edward Rothstein neatly formulated a provocative puzzle raised by the Princeton philosopher Paul Benacerraf a generation ago: "How, if mathematical knowledge stands outside of space and time . . . can [it] be reached from an earthly realm deeply submerged in space and time?" The human brain comes wired to do simple arithmetic and geometry; the rest is invention and discovery. The language and methods of mathematics are the technology that somehow equip earthbound intellects to journey through the universe of mathematical knowledge, so it should not be surprising that it lies uneasily on human tongues.

After Erdős's father convinced him that the primes go on forever, he showed him another beautiful proof that would be the spur for his lifelong obsession with prime numbers and inspire some of his most famous results. *Apuka* asked: How far apart can two successive primes be? Since all primes greater than 2 are odd (2 slips through since, by definition, a prime is a number divisible by only 1 and itself), the difference between two successive primes greater than 2 must be an even number (test this by looking at some small primes). How large can this even number be? Using a method very similar to Euclid's proof, Erdős's father showed him how he could find stretches of integers of any length that were entirely free of primes.

Say, for example, you wanted to find a run of 100 composite—that is, nonprime—integers. First multiply together all the numbers from 1 to 101—that is, $1 \times 2 \times 3 \ldots 101$. Mathematicians call this number 101 factorial and write it as 101! The exclamation mark is standard notation as well as an acknowledgment of the fact that factorials are mostly astonishingly huge numbers; 101! is, approximately, 9.4×10^{159}, a number 160 digits long. The useful thing about 101! is that it is a multiple of every integer from 1 to 101. Since 101! is a multiple of 2, adding 2 to it results in another multiple of 2. Add 3 to 101!, which is also a multiple of 3, and you get another multiple of 3. And so on up to 101. Therefore, none of the 100 numbers from 101! + 2 to 101! + 101 can be prime. This trick—mathematicians would call it a construction—can be used to find arbitrarily long stretches of composite integers. Want a thousand consecutive composite numbers? Easy. Start with 1,001! + 2 and end with 1,001! + 1,001. But don't bother trying to write these numbers down since they are each 2,568 digits long.

Primes can also be very close together. Primes like 29 and 31, separated by only a single number, are known as twin primes. A famous unanswered question is whether or not there are an infinite number of twin primes. The largest known twin primes, which were found in 1994 with the help of supercomputers, are 4,932 digits long —$697,053,813 \times 2^{16,352} \pm 1$. "It is *clear* that there are infinitely many but nobody will prove it in the near future, I think," Erdős liked to say when he lectured on difficult problems. Conviction, however, is not proof. The twin prime conjecture has resisted all attempts at proof for more than two thousand years, and most mathematicians do not feel that a proof will occur any time soon.

Taken together, the facts that primes can be arbitrarily far apart and yet twin primes probably go on forever show how truly strange is the distribution of prime numbers. No magic formula exists that can tell exactly which numbers are prime and which are not. The only surefire way of determining whether a number is prime is laboriously to divide it by every smaller prime number.* Gauss maintained that the problem of distinguishing primes from nonprimes was "one of the most important and useful in arithmetic. . . . The dignity of the sci-

*Actually, you only have to test primes that are smaller than the square root of the number, a large savings of labor. When a number is divided by another that is greater than its square root, the result is a number smaller than the square root. For example, the square root of 36 is 6. Dividing 36 by 2, a number smaller than 6, gives 18, a number that is larger than the square root. To prove that 37 is prime it is only necessary to divide it by primes less than 6, since if it had a prime factor greater than 6, it would have to have one less than 6 as well.

ence itself seems to require that every possible means be explored for the solution of a problem so elegant and so celebrated."

Hearing these proofs and conjectures about prime numbers from *Apuka* inspired in the ten-year-old Paul Erdős an obsession with prime numbers and their distribution that would last a lifetime and would lead to some of the most beautiful and unexpected mathematical results of the twentieth century. The first such result was only seven years in coming. The most important result of *Apuka*'s lesson, however, was that it caused Paul to realize that he would be a mathematician, though he was also fascinated by the stars and thought he might like to be an astronomer.

At the age of eleven, for the first time, Erdős enrolled in a school, the Gimnásium of Tavaszmezö Street, in the sixth grade. Whatever distress his independent spirit may have suffered under the discipline and tight restrictions of the classroom, Erdős nevertheless effortlessly managed to reach the top of his class; the only subject in which he did not receive an A was drawing, for which he got a B. "To learn something was an easy task for me," he would recall. History was his favorite subject and would remain a lifelong interest.

For Erdős, mathematics was more than just another school subject. The year he started attending classes at the gymnasium was also the year Erdős first laid eyes on the pages of *Középiskolai Matematikai Lapok* (the Mathematical Journal for Secondary Schools), known as *KöMal* for short. *KöMal* was created in 1894 by a brilliant young mathematics teacher from Györ named Dániel Arany, as he wrote, to "give to teachers and to students an ample collection of mathematical exercises." Within three years the journal was taken over by László Rátz, the "miraculous" teacher of eleven-year-old John von Neumann and twelve-year-old Eugene Wigner. Under Rátz's direction *KöMal* evolved from something more than a "collection of mathematical exercises" to an immensely successful breeding ground for mathematical talent.

Every month the pages of *KöMal* contained articles by well-known mathematicians and educators of the day. But as interesting as those were, the real draw was the contest problems. Upon receipt of each month's *KöMal* talented math students all over Hungary would go to work concocting the most beautiful solutions they could to these cleverly constructed problems, which were tailored in difficulty to each age group. The solvers would mail their solutions back to *KöMal*, where a panel of volunteer judges would grade them. The very best solutions were published in *KöMal*, a practice that may have been the inspiration for Erdős's fantasy about The Book. At the end of the year pictures of the top all-around solvers were printed for all to see. "The main thing was to solve problems," Marta Svéd, one of Erdős's fellow solvers, would remember years later. "The reward was that if your picture appeared at the end of the year as an industrious problem-solver, well, you felt at the top of the world." In this way talented students all over Hungary became known to each other and began to think of themselves as mathematicians who were part of a larger mathematical community. The Nobel prize–winning economist John Harsanyi, the physicists László Tisza and Ferenc Mezei, and the mathematicians George Polya and Gabriel Szegö were some of the distinguished scientists who started out as top *KöMal* problem-solvers.

Forty years after Erdős picked up a copy of *KöMal*, László Lovász, another child prodigy who was a protégé of Erdős, would see his first issue. "It was love at first sight," Lovász recalls. The issue of *KöMal* that the eighth-grader picked up contained an article contributed by Paul Erdős, by then a world-famous mathematician but still devoted to *KöMal*. "I read the article at least twenty times," Lovász wrote in the centennial issue of *KöMal*. "I was surprised and excited to learn that I could understand what great mathematicians think about." Lovász's love of *KöMal* was shared by many of its readers. George Szekenes, forced to flee Hungary during the war, managed to carry his bulky collection of back issues with him first to Singapore and then on to Australia, where it today occupies a place of pride on his bookshelves.

Erdős's prolific mathematical career can be said to have started with the publication of his solutions in *KöMal*, the first of which appeared in December 1926. Erdős's first joint publication in his long collaborative career occurred when he and a student he had not yet met, Paul Turán, were the only two to solve a difficult problem. Turán would become one of Erdős's closest friends and most important collaborators. Erdős's picture first appeared among the top problem-solvers in 1926. In the photograph, which he also thriftily used on his bus pass, Erdős is wearing an open-collar shirt and staring directly into the camera, unsmiling and very serious. He looks very young, even compared with the other solvers, who are for the most part more formally dressed, with starched collars and neckties, though no less serious. All but one of his fellow problem-solvers that year were men. The exception was a young woman with short-cropped hair named Esther Klein, who would become another of Erdős's lifelong friends and would provide the inspiration for one of his most important and fruitful papers.

Erdős attended the Tavaszmezö gymnasium for two years, then spent another year studying at home before returning to finish his high school studies at the Szent István Gimnásium, where his father taught math and physics. In 1920, in the wake of the rising anti-Semitism that followed the collapse of the Hungarian Commune and the dismemberment of Hungary by the Trianon peace treaty, an act known as the *Numerus Clausus* limited Jewish admissions to the university to 6 percent. Jewish university students were frequently beaten. "At twelve I knew that eventually I'd have to leave Hungary because I am a Jew," Erdős later remarked. Meanwhile, despite the *Numerus Clausus* and violent anti-Semitism, upon graduating from Szent István's Erdős entered the Pázmány Péter Tudományegyetem (the Science University of Budapest).

At the university Erdős soon became the focus of a group of a dozen or so young mathematicians. Most he knew only from their shadowy portraits in *KöMal*. Now, meeting them for the first time in the flesh, he began the mathematical conversations and friendships that would occupy the rest of his life. The students' discussions quickly spilled out of the classrooms onto the streets and cafés of Pest and on long walks through the beautiful wooded hills of Buda, where they learned to do math without pencil or paper. But their favorite meeting place was at the feet of a statue in Budapest's City Park.

The City Park, in the middle of Pest, is where many of Budapest's residents enjoy strolling away an idle afternoon. It lies at the end of the geometrically straight and broad Andrássy Avenue, which was modeled on the Champs-Elysées. The Bois de Boulogne was the inspiration for the City Park, which, along with Europe's first subway line, which runs beneath the Andrássy to the park, was built as part of Hungary's millennial celebration in 1896. The Andrássy and the City Park both fall slightly short of their French originals; the former was not quite as wide and the latter was a bit dusty. Nevertheless, on summer afternoons visitors enjoyed boating on the lake, visiting the zoo and circus, or strolling around the ornamental castle. On winter days there was ice-skating on the lake. Summer or winter, in the decades after World War I, if you knew where to look, you could also join with a group of young men and women engaged in the serious fun of mathematics.

The Vajdahunyad Castle in the City Park is a rambling confection of a building, constructed to embody in a single structure an encyclopedia of Hungarian architectural styles. In the middle of its cobbled courtyard is a large bronze statue of a seated man. His face is entirely hidden by the cowl of his heavy robe, but his gaze is clearly fixed on the pages of a large book in his lap in which he is assiduously writing an invisible text. The figure represents Anonymous, a shadowy medieval chronicler of Hungarian history. The Statue of Anonymous is an ideal rendezvous: easy to find, yet off the beaten path; shaded by leafy trees, and with a bench around its base. Once or twice a week in the early 1930s Erdős would walk from his apartment not far from the park to the statue to meet with his group of university friends. They talked about many things, but mostly the talk was of mathematics. Sitting there, bent over notebooks open on their laps, they echoed in their pose the Statue of Anonymous, which floated above their heads like a dream. When those informal seminars began, neither Erdős nor his friends had yet published a paper in a professional mathematics journal; like Anonymous they were unknown, though not for long.

Even without Erdős, it would have been as remarkable a group of young mathematicians as could be found anywhere. The ten or twelve students who gathered regularly in the park to solve mathematical problems never thought of themselves as a group; that term only came later. Its most famous members included Paul Turán, Tibor Gallai, and George Szekeres, who would all become leading mathematicians in their own right—and Erdős's first collaborators.

The students who met at the statue were all Jewish, though this was something they rarely acknowledged among themselves. Years later, in a conversation about the old days with Erdős, Andrew Vazsonyi mentioned that he felt "there was a wall between the Gentiles and the Jews" in Budapest. Erdős claimed to have never noticed, so Vazsonyi asked him to name some non-Jewish friends from that era. He couldn't name one. "I never thought of that," he admitted.

Shortly after he began attending the university, Erdős made his first significant contribution to mathematics. Ever since his father had taught him about the quirky distribution of prime numbers, Erdős had been drawn to number theory, the branch of mathematics concerned with the properties of the integers. In this he was hardly unique, since most mathematicians are first drawn to the subject by the beauty of the results of number theory and the allure of the questions it poses. Most other branches of mathematics are inhospitable to outsiders; it can take years to understand questions in, say, algebraic topology or group theory, and even longer to master the technical machinery needed to answer such questions. The questions that interest number theorists, on the other hand, can be understood by anyone who has mastered simple arithmetic, though the answers are often exceedingly difficult. A good example is Bertrand's Postulate, the problem that formed the basis of Erdős's first paper.

When Jules Verne was puzzled about some scientific issue he would frequently seek out Joseph Bertrand, a mathematician and member of the French Academy of Sciences. Bertrand did important work on the orbits of celestial bodies and the theory of probability, but he is best remembered for an inspired guess he made in 1845. The twenty-three-year-old mathematician postulated that there was always at least one prime number between every number and its double. A quick check proves that Bertrand's Postulate is true for small numbers. The prime 5 lies between 3 and 6; the primes 17, 19, 23, and 29 lie between 15 and 30. Using the tables of his day Bertrand could have checked his hypothesis for numbers into the millions. Such a check would have been reassuring, but for Bertrand's Postulate to be elevated to the more revered status of a theorem required a proof that he was unable to provide. Five years later a proof of Bertrand's

Postulate was found by Pafnuty Chebychev, a Russian mathematician; it is now called Chebychev's theorem.*

Chebychev's proof of Bertrand's Postulate was correct, but it was also long and difficult. Sometimes difficult proofs are unavoidable, but often they are signs that the essential nature of the problem has not been understood, or that a deeper insight was lacking. Cleaning up old proofs is an important part of the mathematical enterprise that often yields new insights that can be used to solve new problems and build more beautiful and encompassing theories. "Leafing through any of the three thousand–odd journals that publish original mathematical research, one soon discovers that few published research papers present solutions of as yet unsolved problems," the mathematician Gian-Carlo Rota has observed. "The overwhelming majority of research papers in mathematics is concerned not with proving, but with reproving; not with axiomatizing, but with reaxiomatizing; not with inventing, but with unifying and streamlining; in short, with what Thomas Kuhn calls 'tidying up.'" For eighty years it seemed that Chebychev's proof was difficult because it had to be. Then the teenage Paul Erdős saw how the proof could be tidied up.

Erdős's new proof of Chebychev's theorem is simple enough to be understood by an undergraduate. Erdős gave a seminar on his proof at the university to an audience that should have been able to understand the proof. Nobody did. Erdős was never a great lecturer, at least in the usual pedagogical sense, though in small groups or one to one he was a brilliant teacher. In later years Erdős lectured frequently to large and appreciative crowds. By then he had evolved a *shtick*—a mixture of jokes, anecdotes, and mathematics that led one colleague to dub him "the Bob Hope of mathematicians." But in his university days Erdős's act was not yet ready for the road; the only thing that was clear to those who attended the seminar was that Erdős thought he had a neat proof of Chebychev's theorem. Erdős's writing style was no better; nobody could understand the paper he wrote describing his proof, either. "It is one thing to prove a mathe-

* Other similar conjectures about the distribution of primes are even more difficult to prove. For example, is there always a prime between two consecutive square numbers, such as between 8^2 and 9^2? Apparently yes, but so far nobody has been able to prove it.

matical theorem and another to formulate it so that colleagues can comprehend it," Erdős admitted. "Then I was still inexperienced."

Erdős's father could not tell if his son's claim of proving Chebychev's theorem was right, but he knew someone who could. He persuaded László Kálmár, a professor at the University of Szeged and one of Hungary's leading mathematicians, Andrew Vazsonyi recalls, to "take a day off and try to read Erdős's mind." Kálmár struggled with the paper all day and finally emerged from his office at 3 P.M., convinced that Erdős was, as Marta Svéd, one of Erdős's friends from the Anonymous group, would say, "the real thing." Kálmár rewrote the paper in German and had it published in the *Acta* of Szeged, a small mathematical journal. Kálmár's name did not appear as coauthor of the paper, but this method of working prefigured all of Erdős's future collaborations. As one of his collaborators, Arthur H. Stone, would describe it, "Paul's method of writing joint papers was, of course, for him to convey just the essence of his share of the argument; it was up to the co-author to write the actual paper. (I believe Alexandre Dumas père used a somewhat similar system.)"

In his introduction to Erdős's paper Kálmár mentioned that Erdős's proof somewhat resembled a previous proof by the great Indian mathematician Srinivasa Ramanujan, a connection that would become a vital source of inspiration for Erdős. Ramanujan's life is one of the most fascinating rags-to-riches stories in the history of mathematics. He was born in 1887 to a poor family from a town outside Madras in south India, a region viewed by the sophisticates to the north as backward and superstitious. It was a far cry from Budapest, which provided everything needed to nurture the genius of a Paul Erdős. Ramanujan taught himself mathematics from outdated textbooks and invented whatever he didn't know. Finally he wrote a letter to the leading British mathematician of his day, G. H. Hardy, describing some of his theorems and asking for advice. The letter began:

Dear Sir,

I beg to introduce myself to you as a clerk in the Accounts Department of the Port Trust Office at Madras on a salary of only £20 per annum. I am now about 23 years of age. I have had no University education but have undergone the ordinary school course. After leaving school I have been employing the spare time at my disposal to work at Mathematics. I have not trodden through the conventional regular course, which is followed in a University course, but I am striking out a new path for myself. I have made a special investigation of divergent series in general and the results I get are termed by the local mathematicians as "startling."

If Hardy was amused by the mixture of the humble and the arrogant in this letter from a poor, unknown clerk from the middle of nowhere, the ten pages that followed stunned him. "For Hardy, Ramanujan's pages of theorems were like an alien forest whose trees were familiar enough to call trees, yet so strange they seemed to have come from another planet," Robert Kanigel writes in his superb biography of Ramanujan, *The Man Who Knew Infinity*. Like almost all working mathematicians, Hardy was used to receiving mail from cranks, people who were convinced they had figured out how to trisect an angle or square the circle or do something else that mathematicians had long since proved to be impossible. Ramanujan's claims, written in a large, even, and legible hand, were "startling" but not patently impossible. His equations *looked* right.

Writing down complex equations that have the look of real mathematics is not easy; blackboards in old science fiction movies are often filled with scrawlings that most mathematicians would dismiss as trivial nonsense, even at a quick glance. When Hardy finally took the time to go through Ramanujan's equations, he was awed, as his friend C. P. Snow wrote, by the "wild theorems. Theorems such as he had never seen before, nor imagined." Hardy summoned his colleague and chief collaborator, J. E. Littlewood, to his rooms at Trinity College, Cambridge, to help him work through Ramanujan's strange theorems. By midnight the two mathematicians were convinced, as Kanigel writes, that "they had been rummaging through the papers of a mathematical genius." Within a short while Hardy and Littlewood had raised some money and had brought Ramanujan to Cambridge, where in the brief years remaining to him—he died of tuberculosis at thirty-three—he produced a small collection of historic papers and filled notebooks with results that are still being digested today.

Kálmár suggested that Erdős look up Ramanujan's proof of Chebychev's theorem. Erdős went to the library, where he found the proof "in the Collected Works of Ramanujan which I immediately read with great interest.... The two proofs were very similar; my proof had perhaps the advantage of being more arithmetical." The experience of reading Ramanujan's papers would lead Erdős to a life-long interest in India and support of Indian mathematicians. It would also lead to a favorite joke that united his two obsessions: "I think Hindi is the best language because the two greatest evils sound almost the same ... old age and stupidity. Buda is old and Budu is stupid."

Erdős's proof of Chebychev's theorem was a first-rate achievement, especially for an eighteen-year-old university freshman, but, as Kálmár noticed, he had been scooped by Ramanujan. Erdős's proof, however, contained ideas that he quickly applied to generalize Bertrand's hypothesis in an entirely new way. Erdős soon showed that between any number greater than 7 and its double there were always at least two primes. Amazingly, Erdős's proof also tells us something about those primes. According to the theorem there will always be one prime which, when divided by 4, leaves a remainder of 3, and another which, when divided by 4, leaves a remainder of 1. For example, between 10 and 10 doubled, or 20, are the primes 11, 13, 17, and 19. Dividing either 11 or 19 by 4 leaves a remainder of 3, while dividing either 13 or 17 by 4 leaves a remainder of 1. This and a few related results were enough for a Ph.D. dissertation, which Erdős wrote as a second-year undergraduate. They also earned him immortality in a jingle by the mathematician Nathan Fine:

> Chebychev said it, and I'll say it again,
> There's always a prime between N and 2N!

Erdős's work on Chebychev's theorem attracted the attention of mathematicians outside of Hungary. He soon began important correspondences with several of them, including Louis Mordell, a great number theorist in Manchester, Richard Rado and Harold Davenport at Cambridge, and Issai Schur in Berlin. Schur was quick to recognize Erdős's genius and began lecturing to his classes about Erdős's proof of the Chebychev theorem before Erdős was twenty.

Schur's estimation of Erdős's talent increased after Erdős proved a conjecture by Schur concerning abundant numbers. Pythagoras first noted abundant (or excessive) numbers in his search for numerical perfection. Perfection, according to Pythagoras, was reflected in a number's divisors. Pythagoras considered a number perfect if it was equal to the sum of all its divisors less than the number itself. Six is a perfect number since it has the divisors 1, 2, and 3, which add up to 6. Euclid proved that Mersenne primes—primes of the form $2^p - 1$—are related to even perfect numbers by a simple formula. Nobody knows if there are any odd perfect numbers, but the smart money says there are not.

Numbers whose divisors add up to less than themselves are called deficient (or defective) and those whose divisors add up to more are called abundant (or excessive). All primes are deficient, since their only divisor other than themselves is one. Nine is also deficient, since its divisors, 1 and 3, add up to 4. Twelve is abundant, since its divisors, 1, 2, 3, 4, and 6, add up to 16. You can easily verify that 18 is also abundant.

Schur's conjecture concerned how densely the abundant numbers were spread out among the integers. The concept of density is used by mathematicians to compare the size of sets of numbers, even when these sets are infinitely large. The even numbers, for example, have a density of one half since exactly half the numbers are even; multiples of 5 have a density of one fifth. Although the number of primes is infinite the density of primes is zero! That is, the primes are spread so thinly among the natural numbers that the probability that a randomly chosen number is prime is vanishingly small; most numbers are not prime.* Schur conjectured that the density of abundant numbers is greater than zero. Or, loosely stated, that the abundant numbers are indeed abundant.

Erdős found a beautiful, elementary proof of Schur's conjecture. This proof, as good proofs do, led him to further important work on what are called additive arithmetical functions. In his abundant free time Erdős also solved another of Schur's problems, which caused the German to call the twenty-year-old *der Zauberer von Budapest*—the Magician from Budapest.

---

* As we shall see later in our discussion of Cantor, two sets with different densities can be the same size.

CHAPTER FOUR

THE HAPPY END PROBLEM

‘‘W E mathematicians are all a bit crazy," the German number theorist Edmund Landau told Erdős when they met at Cambridge University in 1935. It was an observation with which Erdős, young as he was, could not but agree, though he found it more amusing than troubling. Crazy or sane made little difference to Erdős; from the beginning of his mathematical journeys he went to great lengths to meet anyone who could produce beautiful proofs and conjectures, and he was very difficult to put off.

Three years earlier one of Erdős's university friends, Sándor Kemény, who worked for the Generali Insurance Company, introduced him to a co-worker named S. Sidon. "He was a good mathematician," Erdős recalled, "but a bit crazier than the average. In fact he was a borderline schizophrenic." Sidon was so shy that he usually spoke facing the wall, "but when he talked mathematics he talked sense." Erdős was impressed by Sidon's sensible observations and felt certain that his friend Paul Turán would be too. So, as he would do so frequently and fruitfully throughout his long life, Erdős played the mathematical matchmaker. Sidon, unfortunately, was a reluctant client.

The two Pauls—Erdős, nineteen, and Turán, twenty-two—showed up at the reclusive Sidon's doorstep unannounced and knocked. The door opened a crack and Sidon peered suspiciously out at his two young visitors, who explained why they had come. "Please visit another time and *especially* another person," Sidon said. "It sounds better in Hungarian," Erdős would point out: *kérem, jöjjenek máskor és különösen máshoz*, which has a snappier rhythm when drawled with the proper accent.

Erdős was not put off, and eventually Sidon "was again reasonable" and invited the young mathematicians in for a chat. The result of those conversations was the creation of a theory of what are now known as Sidon sets, which Erdős would develop in one of his early collaborative papers with Turán; the solitary Sidon, evidently, was one of the few mathematicians Erdős was unable to coax into collaboration.

During their first meeting Sidon asked whether Erdős could show that a sequence of integers having a set of peculiar properties he described could actually exist. "I told him, that's a very nice question, I think you are right, and I hope to give you the answer in a few days," the nineteen-year-old Erdős boasted. He later admitted that he had been "a little bit too optimistic. I eventually did prove it but it took twenty years." Sidon died a few years later, and so did not live to see Erdős's bravado justified. "Actually his was a curious death," Erdős would tell lecture audiences. "He died like Cyrano de Bergerac: a ladder fell on him and broke his leg, and he died in the hospital of pneumonia." In a typical gesture, Erdős dedicated his belated proof to the memory of Sidon.*

Erdős's restless energy; his ravenous hunger for mathematical con-

---

* Erdős proved the existence of a sequence of integers having the properties specified by Sidon by using one of his greatest inventions, the "probabalistic method." He was unable to construct an example of such a sequence and offered a prize of $300 to anyone who could. The prize is unclaimed, but like all prizes Erdős offered in his lifetime, it will be paid off by a fund established by his friends to anyone who qualifies.

versation, collaboration, and community; and his famous idiosyncrasies were already in evidence during his university years in Budapest. George Szekeres, a few years older than Erdős and his earliest collaborator, recalls that when he first met Erdős he "found him a very puzzling phenomenon. He was utterly, what you would call, neurotic." Once, Szkeres recalls, when Erdős was sitting with his group of friends on the bench at the foot of the Statue of Anonymous, "he suddenly jumped up, without any reason, rushed to and fro, and came back again. There was an old lady who watched all this and called me aside and asked, 'is there anything wrong with this young man?' Nothing was wrong. We were just talking and suddenly something probably occurred in his mind, he jumped up to sort it out and came back. He makes a very queer impression on the outsider. . . . I think, we all did the right thing with him. We behaved totally naturally, we accepted this."

Erdős's parents were popular among his friends, who were frequent visitors to the Erdőses' spacious, modern Budapest apartment at 8 Abonyi Street, across from the Jewish High School. "For students it was an open house," Esther Klein, a frequent visitor, fondly recalled. Erdős's parents were a source of advice, hospitality, and—best of all to Erdős's often cash-strapped friends—the names of students who needed tutoring, of which his parents, as teachers, had a constant supply. At the Erdőses', Paul's friends could also witness at close range his mother's protectiveness. Shortly before Erdős left Hungary for the first time, Szekeres recalls Erdős's mother taking him aside to make a private request. She asked Szekeres to promise that if either she or her husband died when Erdős was abroad that he keep this fact from his friend so as not to upset him and distract him from his studies. "That's what people call overprotective," Szekeres observed.

The protectiveness of Erdős's mother could be touching. Andrew Vazsonyi recalls receiving a letter from Anna Erdős. Like most mothers, she was concerned about her son's future, though she probably had better reasons than most. Erdős had, after all, chosen to pursue a career for which, as a Jew, there was little future in Hungary. Besides, he was impractical, dependent, preoccupied with mathematics, and, worst of all, seemed to be totally unconcerned with the future. "What is to become of my son?" she wrote despairingly. "What is he going to do for money?" Vazsonyi replied that "the ordinary rules of life do not apply to a genius. . . . He will manage somehow. Anyway, he has unlimited credit with his friends."

Most of all *Anyuka* protected Erdős from the clutches of women. One day, while walking with his girlfriend through the City Park, Vazsonyi met Erdős. The couple accompanied Erdős back to his nearby apartment building. Many Budapest apartment buildings were built around a central courtyard that was visible to all from a ring of external balconies. "As we entered the court," Vazsonyi recalls, "a sudden scream reverberated throughout the court, to be heard by all: '*Who is that woman?*' It was Erdős's mother yelling in total panic. 'Just Vazsonyi's girlfriend, Aranka,' said Erdős humbly, and she calmed down. A girlfriend of a friend was safe, [since] she was absolutely out of bounds for all others."

Erdős would never marry, and the question of his sexuality would always remain mysterious to his friends. In George Csicsery's fascinating documentary film about Erdős, *N Is a Number*, Erdős remarks on the subject somewhat obliquely: "As somebody put it, 'he likes girls but he doesn't like the thing which they are standing for.' " Whatever precisely the "thing" was, it did not prevent Erdős from having women as close friends and collaborators. That is amusingly illustrated by a joke mathematicians like to tell: On one of his frequent journeys across the United States Erdős decided for once to ride the train. As luck would have it he found himself seated next to a stunningly beautiful young woman. The two struck up a conversation, and one thing led to another. By the time the train was pulling into Penn Station, they had written a joint paper.

Erdős did have at least one extramathematical relationship with a woman. One day in the 1960s Vazsonyi, who at the time was living and working in Los Angeles, received a call from Erdős. "*Itt vagyok!*" he announced. "I am here." It was how he began all calls. Erdős had just arrived at UCLA and summoned Vazsonyi to the cafeteria. Vazsonyi was used to this kind of thing; Erdős was unable to drive, and when he was in the area Vazsonyi was his chauffeur. But this time Erdős had a surprise. "Vazsonyi, you don't have to drive me around anymore," Erdős announced.

"Erdős, what does this mean?" Vazsonyi asked.

Erdős explained that a woman he had met named Jo Brüning, a Dutch physicist, would now take him wherever he needed to go. "This was a total amazement to me," Vazsonyi recalls, shaking his head. "Every time we went anywhere with Erdős, Jo was there." On Erdős's initiative they visited the California missions, though Jo, a devout Protestant, refused to pay the entry fee to the Catholic missions, and waited stubbornly outside. The two of them also joined Vazsonyi on a trip to Laguna Beach that ended badly. Erdős had reserved separate rooms for Jo and himself, only to discover on checking in that only one room was available. The desk clerk was apologetic, and Vazsyoni suggested the two share a room. Erdős turned around, very annoyed, and shouted, "Impossible, that's just impossible!" Later Jo confided to Vazsyoni's wife, Laura, that she was going to break up with Erdős. "I'm tired of being his chauffeur."

Erdős's relationship with Brüning was an experiment that was almost certainly bound to fail. "I have an abnormality," Erdős explained in Csicsery's film, his voice filled with pain. "I can't stand sexual pleasure." Erdős recoiled from even the slightest physical contact; when a stranger offered a hand to shake the best he could manage was to brush it lightly with his limply extended hand. Even a fleeting, accidental touch made him extremely uncomfortable, and he washed his hands compulsively all day long.

Throughout his life Erdős hated to be alone, and he would surround himself at every opportunity with friends and colleagues. He loved to stay with friends, especially those with young children, who would call him Uncle Paul. But Erdős never stayed anywhere for long and even in a crowd, surrounded by his inviolable personal space, he was deeply alone. He always believed that his "abnormality" made him "almost unique" among humanity. Watching Erdős walk slowly across a park or down a corridor in Csicsery's film, one is struck by his extreme isolation. Erdős saw his excruciating separateness as a fundamental part of his psychological makeup and a source of strength as well as pain. "I have a basic character that I always wanted to be different from other people," he explained. "It's very, very much ingrained. From a very early age I automatically resisted pressure to be like others."

Nobody knows the origin of Erdős's abhorrence of physical contact, though one family member suggests that the problem is due to a congenital medical condition. Whatever the cause, it did not prevent Erdős's mother from occasionally teasing him. Once, at a gathering at Lake Balaton in Hungary, Janos Pach, a friend and collaborator of Erdős, heard *Anyuka* call out, "Paul, why don't you have any children?"

"Mother, you do not think I am impotent, do you?" he replied.

Most of Erdős's male friends—and most, though not all, of the young mathematicians in his circle were male—were as obsessed with women as they were with mathematics. Erdős would not accompany them on their trips to the movies to ogle beautiful American actresses. Vazsonyi recalls that they liked to make Erdős feel uncomfortable by talking graphically about sexual matters. Once some friends came across Erdős making up packages of food to be sent to China to relieve suffering during the Civil War. As a joke they offered to make a hundred-dollar contribution to his cause if he would accompany them to a burlesque show. Knowing of Erdős's sexual squeamishness, they felt sure their money was safe. So they were shocked and concerned when Erdős agreed. After the performance, having happily collected the money, Erdős said with a grin, "See, you trivial things, I tricked you. I took off my glasses and did not see a thing!"

The use of the word "trivial" to mean stupid is an example of the dialect known to all mathematicians as Erdősese, which Erdős began to develop during his university years. Erdősese has etymological roots in borrowings from the jargon of mathematics. For example, Erdős loved children, whom he insisted on calling *epsilons*, after a Greek letter that mathematicians use to denote vanishingly small quantities. When his friend László Alpár was arrested and imprisoned for his participation in Communist student activities in 1933, Erdős informed his friends that "L.A. is studying Jordan's theorem." Jordan's theorem states what may seem obvious but turns out to be extremely difficult to prove, that every closed curve divides the plane into an inside and an outside. Alpár, Erdős was saying, was on the inside. Erdősese can be educational: George Szekeres first learned of this important theorem from Erdős's quip.

Such circumlocutions were meant to be amusing, in an academic sort of way, but they also served a more practical purpose. It could be dangerous to be overheard speaking about political matters. So, in Erdősese, the Soviet Union was always "Joe" for Josef Stalin, and the United States was "Sam." Erdős would entertain epsilons with his own version of a famous nursery rhyme: Sam and Joe went up the hill to fetch a pail of water. . . . If corrected, he would solemnly explain that Jack and Jill were Elizabethan politicians. Communists were on "the long wavelength," since the wavelength of the color red is the longest in the visible spectrum.

Music, which Erdős did not care much for in those days, was "noise." Even when he later came to love music, it remained noise, a term that seemed particularly appropriate to his hosts, who had to bear his playing it on the radio at all hours of the day. Alcohol, which he rarely drank, was "poison." Less politically correct were his terms for human relationships. When a friend married, Erdős said he had been "captured." Wives were "bosses" and husbands "slaves." He could never understand why this irritated some of his friends, and indeed none of his many female collaborators ever noticed the slightest degree of sexism in him. The explanation for those terms is probably simpler: Hungarian wives traditionally called their husbands "master," which Erdős simply inverted.

Perhaps Erdős's most interesting coinage is the term "Supreme Fascist," or SF, which is what he called the God in whom he professed no belief. Erdős's view was that the relationship between human beings and the SF is fundamentally an unfair game that we cannot win but are obliged to play. The SF is the author of The Book of all the best mathematical proofs, and it is one measure of His cruelty that He keeps its contents hidden. We are therefore obligated to use all of our intelligence and intuition to reproduce the contents of the SF's hidden Book for ourselves. When asked, "what is the purpose of life?" Erdős would reply: "To prove and conjecture and to keep the SF's score low." He imagined that humans were constantly engaged in a deadly serious game with the SF in which, "if you do something bad the SF gets at least two points. If you don't do something good which you could have done, the SF gets at least one point. And if you are okay nobody gets any points." Humanity cannot win this game, so the goal of life is not victory. "The aim is to keep the SF's score low."

As the world raced toward another cataclysmic war, Erdős's vision of a universe ruled by an SF must have seemed reasonable. In those irrational times he and his friends took refuge in the rational domain of mathematics. Their frequent excursions to the Buda hills and meetings in the City Park were beginning to move on from practice problems and the rehashing of well-known results to original work and collaborations. George Szekeres would become Erdős's first collaborator, edging out Paul Turán by a few months. Turán's collaboration with Erdős would continue for the rest of Turán's life and would result in thirty papers.

Andrew Vazsonyi was another early, if somewhat reluctant, collaborator. Vazsonyi had been working on an extension of a celebrated problem known as the Königsberg Bridge problem, to which most mathematicians trace the origin of the field of graph theory. Through the Prussian town of Königsberg (now known as Kaliningrad) runs the river Pregel. An island called Kneiphoff lies in the middle of a fork in the Pregel, and a network of seven bridges connects the island and banks of the Pregel.

On balmy evenings the burghers of Königsberg enjoyed strolling across the town's seven bridges, and, as they strolled, the question naturally arose: Is it possible to plan a walk so that you cross each bridge once and only once? The great and prolific Swiss mathematician Leonhard Euler tackled this problem in 1736. Euler is second

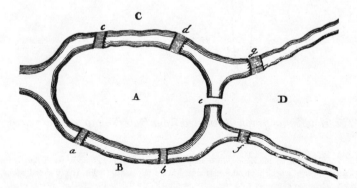

Map of the seven bridges of Königsberg from Euler's 1736 paper.

only to Erdős in mathematical output;* he produced 886 books and papers in his lifetime, more than half of which were written after he became blind at the age of fifty-eight. "I was told that while some deny the possibility of doing this and others were in doubt, there were none who maintained that it was actually possible," Euler wrote. He then set about proving the townspeople correct. Since much of Erdős's life was taken up with graph theoretical problems, it is worthwhile to take a moment to consider Euler's simple proof.

The first thing Euler did was to translate the problem into a form more easily analyzed mathematically by stripping the map of Königsberg of unnecessary elements. He reduced all land masses to points, which a modern graph theorist would call *vertices*. The bridges connecting the land masses became lines, or *edges* in graph theoretical parlance. The result was an abstract representation of the problem as a graph:

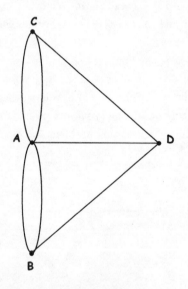

The underlying graph abstracted from Euler's map of Königsberg.

---

* Though by some counts Erdős wrote more mathematical papers than Euler, Euler wrote a vast number of works in physics, astronomy, and other related fields. His collected works, which extend over seventy volumes, far exceed Erdős's published works. On the other hand, Erdős wrote thousands of mathematical letters a year, and if those are ever published their volume would certainly challenge Euler's record.

Euler made the important observation that unless a vertex, which corresponds to a body of land, is the end or the beginning of the walk, it must have an even number of vertices emanating from it. The reason is simple: Since each edge—that is, each bridge—may be crossed only once, every edge entering a vertex must have a corresponding exiting edge. Therefore, vertices with an odd number of edges incident on them—vertices of odd *degree*—must lie at either the end of the walk or the beginning. Now look at the diagram. You can quickly confirm that all four vertices have an odd degree (vertices C, B and D have degree 3 and vertex A has degree 5). Since a walk can have only one beginning and one end, Euler could confidently assert the impossibility of the Königsberger's constitutional. Being a true mathematician Euler went on to pose and solve a more generalized version of the bridge problem, demonstrating the power of mathematical abstraction. "On the basis of the above I formulated the following very general problem for myself," Euler wrote. "Given any configuration of the river and the branches into which it may divide, as well as any number of bridges, to determine whether or not it is possible to cross each bridge exactly once."

Almost exactly two hundred years later Vazsonyi was attempting to generalize the Königsberg problem still further to graphs having an *infinite* number of edges and vertices. Euler failed to make this generalization, perhaps because the concept of infinity was given a precise mathematical form only toward the end of the nineteenth century by Georg Cantor. At their first meeting Erdős had introduced Vazsonyi to Cantor's theory of infinity, which so enchanted him that in homage to the great Cantor Erdős would sign his letters, "C. be with you." Vazsonyi had solved half of the infinite Königsberg—he had discovered the conditions *necessary* for the walk, but not those that were *sufficient*. "I used to meet with Erdős essentially daily and made the fatal error of telling him on the phone about my discovery," Vazsonyi recalls. "I say fatal because he called me back in twenty minutes and told me the proof of sufficient condition. 'Damn it,' I thought, 'now I have to write a joint paper with him.'" Vazsonyi was certain that, given just a little more time, he could have found the solution for himself. In retrospect, however, he is glad he had not, because in the world of mathematics an aristocratic cachet adheres to those who have written a paper with Erdős. But, even after Erdős had become famous, the quickness of his mind and his avidity to solve problems would sometimes cause resentment among mathematicians who preferred to work alone. One case, as we will see later, created a controversy that would divide the mathematics community and scar Erdős's career.

Writing papers with Erdős could have consequences that transcended the world of mathematics. In 1935 Turán wrote a paper on a problem in number theory with Erdős, which appeared in the Russian *Bulletin of the Institute of Mathematics and Mechanics of Tomsk*. Ten years later, in postwar Budapest, Turán was stopped by a Soviet patrol. Stalin had ordered his liberators to round up men randomly in the streets to be shipped to the gulag as *malenkii rabot*, or small labor. The soldier demanded that Turán produce his papers. Turán had lost his ID a few days earlier evading another roundup, but reaching into his briefcase he found a copy of the paper he had written with Erdős. Turán handed the paper to the soldier, who, duly impressed that Turán had been published in a Soviet journal, let him go. Turán later dryly reported this story to Erdős as "a surprising application of number theory."

A simple mathematical puzzle invented by Esther Klein led to another unexpected transmathematical result that would change her life and that of George Szekeres. "I actually remember the moment," Esther said, sixty years after she made her discovery. "I was sitting at home—we had a very simple flat, me and my parents. I had one place where I could sit and think and do some mathematics." She had been filling a pad with geometrical figures—random points that she would connect with lines—when she noticed something that had somehow eluded geometers for more than two thousand years.

To understand Esther's discovery we shall have to pause for a few very simple definitions. A polygon, of course, is a closed, flat figure consisting of straight lines—think of an irregular field enclosed by a fence. Mathematicians classify polygons into two broad classes: those that are convex, and those that are not. A convex polygon is simply a polygon that has no "dents." Mathematically, it can be defined more precisely in two equivalent ways. The first is to say that the angle formed by any two adjacent sides of a convex polygon is less than

180 degrees, measured from the inside of the figure, which is a fancy way of saying "no dents." A convex polygon can also be defined without mentioning angles: The line joining any two points that lie in the interior of a convex polygon always lies entirely within the polygon (see Figure 1). If a polygon is *not* convex, then a line connecting some pairs of interior points will cross the sides and leave the polygon (see Figure 2).

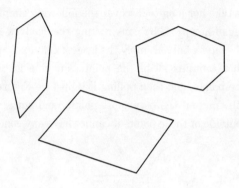

Figure 1
Convex polygons.

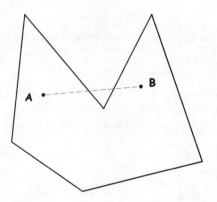

Figure 2
If a polygon is not convex, a line connecting two interior points can cross the sides and travel outside the figure.

• • •

W H A T Esther noticed was that if she drew five points at random on her pad, as long as no three lay on a straight line, four of them would always define the vertices of a convex quadrilateral—that is, a four-sided convex polygon. Always. This odd fact surprised her. As a champion problem-solver she realized that such an assertion required proof; merely drawing example after example would not suffice. It didn't take her long to discover the following simple proof.

Esther began her proof by constructing the "convex hull" of the five points. A convex hull is simply the largest convex figure that can be drawn that contains all of the points. imagine looping a lasso around all the points, and then pulling it tight (see Figures 3 and 4). The rope will snag on the outermost points, and since it is looped around the outside of these points it cannot have any concavities.

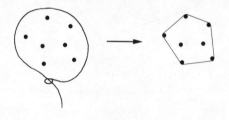

Figure 3
The convex hull can be found by lassoing a set of points.

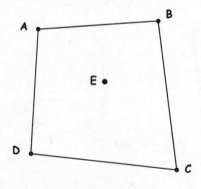

Figure 4
If the convex hull is a quadrilateral, we are done.

In the simplest case the lasso loops around four points. If that happens we are done, because those four points determine a convex quadrilateral (Figure 5).

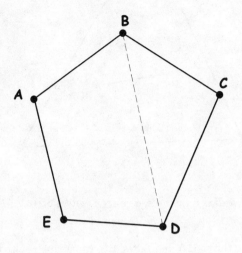

Figure 5
If the convex hull is a pentagon, then connecting two vertices will always result in a convex quadrilateral, ABDE.

The only remaining possibility is that the points were arranged in such a way that the lasso loops about only three of them.* It's not hard to show that in this case as well we can find a convex quadrilateral.

The points might also have been arranged in such a way that the lasso loops all five. In this case all we have to do is join two diagonal points, called B and D in Figure 6. The quadrilateral ABDE is obviously convex.

---

* To avoid ambiguity, no three points are allowed to lie on a line.

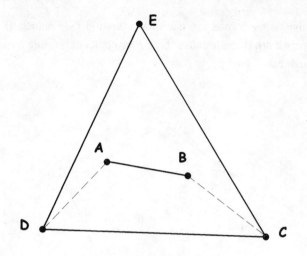

Figure 6
*If the convex hull is a triangle, a convex quadrilateral, ABCD, can
be drawn.*

Draw a line through A and B, the two remaining points inside the
triangle. It's clear that one side of the triangle must lie on one side of
this line. In our diagram these points are D and C. As you can see,
the quadrilateral ABCD is convex.

Quod erat demonstrandum. We are finished. We have analyzed
every possible configuration of five points and shown how each one
must contain a convex quadrilateral.

To Esther the proof was "childishly simple." What was interesting,
she realized, was that "it was a new sort of question. I think that was
its real merit." Esther's puzzle combined geometry with combinatorics. Combinatorics is the branch of mathematics concerned with
counting things. Erdős was perhaps the most important and prolific
combinatorialist of the twentieth century, and a thick collection of his
classic papers on the subject is called *The Art of Counting.* Before
Erdős the subject of combinatorics consisted of a relatively small
number of unrelated techniques and problems. Great mathematicians
dabbled in combinatorics but devoted most of their energy to other
fields. In this century, thanks largely to the work of Erdős, the subject has emerged as a field in its own right, with its own textbooks

and journals and international congresses. Combinatorics is also important to the design of communications networks and computers and has applications in almost every branch of science and technology. Erdős's influence on these fields has been immense, though indirect. For example, there have been conferences on Erdős's interest in computer science, though he never wrote a paper about computers, and he was a frequent and valued visitor to the mathematics division at Bell Labs, though he never wrote a paper about communications networks. To Erdős doing combinatorics, like all of his other interests, was doing math for math's sake. And his interest in the field started with Esther's question.

Esther took her puzzle to the Statue of Anonymous to show her friends. They rapidly solved it and moved immediately to generalization. After a little more effort they proved that a convex pentagon is unavoidable when nine points are strewn willy-nilly across the plane. Were convex polygons of six, seven, eight, or any other number of sides also inevitable given enough points? If so, exactly how many points would be needed in each instance? Those were more difficult questions, and it was clear that no "childish" solution would do. Both Erdős and George Szekeres were immediately fascinated by the new question. They saw, as Esther had, that it was a new sort of problem. But Szekeres recalls that he had another, deeper motivation that had nothing to do with geometry or combinatorics. "I had no other feeling but that I wanted to solve the problem because it came from Esther," he later recalled. Szekeres believes that Erdős's interest in the problem was also inspired by his deeper and more human interest in its creator. "I am sure if anyone had suggested this to him he would have violently objected, but that doesn't mean it's entirely untrue," Szekeres said.

Erdős had always been especially fond of Esther, whom he affectionately called Epszi, short for the useful mathematical diminutive *epsilon*. More importantly, Erdős's mother approved of Esther and welcomed her into her home. "I'm quite sure that old mom would have been very happy if Paul married Esther," Szekeres speculates. "To me it was quite obvious that Esther would have had the greatest chance to make him marry. It's very complicated. Probably the world is better off like this because he could really devote himself to mathematics."

When Erdős wanted Szekeres to listen to a new proof or a conjecture, he would intone, "Szekeres Gy., open up your wise mind," employing the form of his name, Gyorgy, that Szekeres used to sign his mathematical papers. Two weeks after Esther had posed her generalized problem, Szekeres had the pleasure of turning the tables and demanding of Erdős, "E.P., open up your wise mind!" He had found an ingenious proof of their conjecture that a convex polygon with any number of specified sides is inevitable when a sufficient number of points are randomly strewn on the plane.

To prove their conjecture Szekeres had unknowingly rediscovered a theorem that had been published in 1928 by a brilliant young British polymath named Frank Plumpton Ramsey. Ramsey was born in 1903 and raised in Cambridge, England, the son of the mathematician Arthur S. Ramsey, who was the president of Magdalene College; Frank's younger brother was the Archbishop of Canterbury. In his brief life—he would die one month shy of his twenty-seventh birthday of a chronic liver disorder—Ramsey would write only a handful of papers in mathematics, philosophy, and economics. Most of them would become classics.

Like Erdős, Ramsey was educated at home by his mother. After a few years of home schooling Ramsey went off to the Winchester Public School, where his brilliance quickly became apparent. One day Ramsey announced to his friends that he wished to learn German. He took home a grammar and a dictionary, and within a few weeks could read and critique the Austrian physicist and philosopher Ernst Mach's book, *Analysis of Sensations*, in the original.

Ramsey raced through Winchester and then enrolled at Trinity College, Cambridge. He would eventually receive first class honors in mathematics, but Ramsey's restless mind would not be confined to any one discipline. When he was just sixteen Cambridge economists flocked to have their ideas subjected to Ramsey's keen scrutiny. John Maynard Keynes wrote that Ramsey handled "the technical apparatus of our science with the easy grasp of one accustomed to something far more difficult."

Nothing seemed too difficult for Ramsey. Keynes called the second of the two papers Ramsey wrote on economics "one of the most remarkable contributions to mathematical economics ever made." Ramsey also wrote important papers about Wittgenstein's logical paradoxes and an original interpretation of the theory of probability. But his most important work, which gave rise to the theorem that was rediscovered by Szekeres, was on mathematical logic. Ironically, Ramsey's theorem, which would become his principal claim to fame, was, as Ronald Graham and Joel Spencer wrote in a *Scientific American* article on Ramsey's theory, "superfluous to an argument which he could never have proved in the general case."

The subject of Ramsey's paper was an attempt to deal with some fundamental issues put forth by Alfred North Whitehead and Bertrand Russell in their monumental work, *Principia Mathematica*. Taking their cue from Euclidean geometry, Whitehead and Russell tried to show in their book that all of mathematics could be derived from a fixed set of axioms and simple logical rules. The German mathematician David Hilbert took this idea a step further and conjectured that one could, at least in principle, find a procedure that could automatically decide the truth or falsity of any mathematical statement. In other words, a computer could be used to determine whether any mathematical statement was true. The key here is in the words "in principle." Mathematicians need not fear for their livelihoods, since such a computer, if it existed, would mostly provide proofs that would be anything but the elegant Book proofs that illuminate; computer proofs of even the simplest theorems would be extremely long and ugly, and offer no guidance or insight. Russell's and Whitehead's formal proof of the equation $1 + 1 = 2$ that appears in the *Principia* is famously long and obscure, the kind of thing that is read only by specialists and other masochists. A computer following Hilbert's hoped-for procedure might need thousands of years to decide whether a proposition were true or false, but the hope that a proof—any proof—must exist was a vital element of his mathematical credo. "We hear within us the perpetual call. There is the problem. Seek its solution. You can find it by pure reason, for in mathematics there is no *ignorabimus*," no remaining ignorant.

A few years later, the British mathematician Alan Turing, expanding on the pioneering work of Kurt Gödel, shattered Hilbert's belief by proving that, even in principle, it is impossible to program a machine that could decide the truth or falsity of all statements. In a paper whose unsettling repercussions are still being felt in the realms of mathematics and philosophy, Gödel proved that there are mathematical statements that are undecidable, that can be neither proved nor disproved.

Ramsey discovered his theorem in an attempt to do what Hilbert proposed and Gödel and Turing would later prove to be impossible. It might have remained unknown had Szekeres not rediscovered it a few years later. Erdős, who even as an undergraduate read every mathematical journal he could get his hands on, soon uncovered Ramsey's paper.

While the mathematical statement of Ramsey's theorem is couched in formalism that is abstract and difficult, its importance can be understood by looking toward the heavens on a clear night. At first the brightly colored stars appear to be scattered at random across the sky. Upon closer examination the stars seem to limn the outlines of shapes: lines and rectangles, pentagons and circles. Ancient stargazers saw those shapes as the shadows of beasts and gods riding across the storied vault of heaven and believed that the arrangement of the stars revealed the work of an underlying hand. Ramsey's theorem offers a more rational explanation. According to Ramsey not only are shapes like those seen in the sky possible, they are *inevitable* whenever the number of randomly arranged stars is large enough. As the American mathematician Theodore S. Motzkin observed, Ramsey's theory proves that complete disorder is impossible.

To help explain Ramsey's theorem to a lay audience, Erdős frequently resorted to a puzzle known as the party problem. Six people are invited to an intimate get-together, some strangers and some friends. Is it true that among the guests there will always be three people who are all friends or three who are all strangers?

There are many ways to solve this question. Perhaps the easiest way is to start by converting it to a question about a graph, much as Euler did when he tackled the Königsberg bridge problem. In this case, represent each partygoer by a vertex or point. If two people know each other, connect them by a solid edge or line; if they are strangers, connect them by a dashed edge. Every vertex is connected to every other vertex by either a solid or a dashed edge. A graph in which every vertex is connected to every other vertex is known as a complete graph.

If three people all know each other the graph will contain a solid triangle; if three are strangers the graph will contain a dashed triangle. The party problem then reduces to the question: *Can the edges of the graph be drawn in such a way that there is no solid or dashed triangle?*

One way to answer this question would be to examine doggedly every possible party graph and see if there were any without either a solid or a dashed triangle. How difficult would that be? The graph contains fifteen edges. (You can determine that either by counting them in the illustration or by noting that each of the six vertices is connected to five others, which gives thirty edges. But this method counts each edge twice—AB and BA are both included—so you must divide by 2, which gives the correct answer of fifteen edges.) Each of the fifteen edges can be either solid or dashed, so the total number of

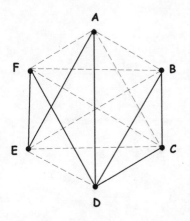

The relationships in the party problem are represented as a graph. Friends are connected by solid lines, strangers by dashed lines. Can a graph be drawn without a solid or a dashed triangle?

graphs is two for the first edge, times 2 for the second, times 2 for the third, and so on up to fifteen. That is, two to the fifteenth power, or 2^{15}, which makes 32,768 possible graphs! A computer could quickly examine all of these graphs, but most people would tear their hair out long before finishing.

With the help of a little logic such auto-depilation can be avoided. Start by focusing your attention on one vertex. To be concrete, label this vertex A.

Vertex A, like all the other vertices, is connected to each of the other vertices by edges that are either solid or dashed. Clearly, at least three of these five edges must be solid *or* at least three edges must be dashed—if the number of solid edges were less than three the number of remaining dashed edges would be more than three, and vice versa. In the diagram below we draw three solid edges emanating from A (the argument would be the same had we assumed the edges were dashed). The remaining edges do not matter.

Remember that we are trying to avoid drawing a solid triangle or a dashed triangle. If any of the three edges, EC, ED, or DC, were solid the result would be a solid triangle. Therefore, none of these edges can be solid: All must be dashed. But these edges themselves

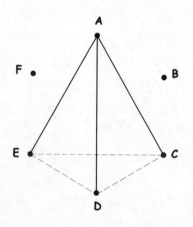

If three of the edges emanating from A are solid (or dashed), avoiding a solid (or dashed) triangle inevitably results in a dashed (or solid) triangle.

form a triangle, ECD! In trying to avoid drawing a solid triangle we are forced to draw a dashed triangle. Had we started with three dashed edges connected to A, we would have been forced to draw a solid triangle. This is exactly what we set out to prove: In a complete graph (party) containing six points (people) there must always be either a solid triangle (three people who know each other) or a dashed triangle (three who are strangers).

Ramsey's theorem generalizes the party problem. It states that as the party gets bigger and bigger there will inevitably be larger and larger groups of guests who either all know each other or are all strangers. These increasingly large knots of inevitable friends or strangers are structures that correspond to the convex polygons in Esther Klein's problem or the constellations in the sky.

As Erdős enjoyed pointing out in his lectures, calculating the Ramsey number—the size of the party—to guarantee a group of three friends or three strangers was reasonably simple, but the problem rapidly gets much more difficult as the group of friends or strangers increases. Mathematicians symbolically designate the size of a party in which there must be either three friends or three strangers $R(3,3)$. Even without any clever insight, a computer could have found $R(3,3)$ by mechanically examining different party sizes until it hit upon the answer, a party of six. That would not have taken long, even on a slow computer, since the number of different parties of six is only 32,768.

What if we asked how large the party would have to be to guarantee that there would always be a group of four mutual friends or four mutual strangers? In other words, what is $R(4,4)$? The answer turns out to be 18, though the proof, says Erdős, "is not quite so simple any more." Now it really does pay to be ingenious, because there are about 1.14×10^{46} possible ways to represent a graph having eighteen points. To give an idea of the staggering magnitude of this number requires ludicrous metaphors, so here's one: If every atom in your body were a high-speed computer, together these computers would not come close to being able to check this solution in the age of the universe. Such is the power of mathematical reasoning, which substitutes reason for brute force.

But the power of mathematical reasoning is soon exceeded by the party problem. Nobody knows exactly how big parties need be before a clique of five mutual friends or strangers is inescapable. More than half a century of work by the brightest mathematicians in the world has narrowed the answer to greater than 42 and less than 50. In his lectures on Ramsey theory Erdős delighted in spinning a fantasy to impress on audiences the astonishingly rapid increase in the difficulty of this problem. "Suppose an evil spirit would tell mankind, either you tell me the answer with five people or I will exterminate the human race. . . . it would be best to try to compute it, both by mathematics and with a computer. If he would ask for six people, the best thing to do would be to destroy him before he destroys us, because we couldn't do it for six people. Now, if we could be so clever that we would have a mathematical proof, we could just tell the evil spirit to go to hell."

While Erdős struggled to read the Ramsey numbers hidden in the Supreme Fascist's Book, others have pored over the pages of another book for concealed messages. Ramsey's theorem and the logic of chance guaranteed that they would succeed. Michael Drosnin, in his best-selling book *The Bible Code*, expanded on the work of Eliyahu Rips of Hebrew University in Jerusalem to find messages buried in the text of the Bible. Using a computer to search the 304,805 letters of the Hebrew Bible, Drosnin claims to have found prophecies and omens enough to evoke the combined envy of Joshua, Nostradamus, and the Amazing Karnak. Drosnin finds his messages by examining letters in the Bible that are spaced a fixed distance apart. Consider, as an example not drawn from Drosnin's book, the phrase from the King James Version of Exodus, 31:28: "And hast not suffered me to kiss my sons and my daughters? thou hast now done foolishly in so doing." Starting with the R in "daughters" skip four letters (ignore spaces and punctuation) to the O in "thou," then four more to the S in "hast," and so on. The result is the word ROSWELL. Starting with the U in "thou," skipping eleven letters gives an F and skipping another eleven gives an O. Thus, UFO is buried in the same passage as ROSWELL, which might be taken by some as proof that the Bible foretold the coming of aliens to the New Mexico desert.

Using a computer to take the tedium out of his searching, Drosnin found many such seemingly significant words and phrases juxtaposed throughout the Bible. For example, Drosnin found the buried word "Dallas" near the word "Kennedy." Using this technique Drosnin claims to have uncovered prophecies concerning almost every fact and figure in human history: the assassination of Yitzhak Rabin, the Gulf War, Adolf Hitler, Bill Clinton, and much, much more.

The "prophecies" hidden in the letters of the Bible are equivalent to the constellations glimpsed by ancients in the desert skies. Although it is impossible to prove that they were not put there by the hand of God or the gods, Ramsey's theorem tells us that, given enough letters or stars, they are inescapable. The physicist and skeptic David Thomas searched the King James Bible, *War and Peace*, and other texts, both sacred and profane, for prophecies. He found as rich a horde in these texts as Drosnin found searching the Hebrew Bible. To proponents of the Bible Code who ask, "How could such amazing coincidences be the product of random chance?" Thomas replies with what he feels is the real question, "How could such coincidences not be the inevitable product of a huge sequence of trials on a large, essentially random database?"

Pictures beamed back by a space probe in the 1970s offer another opportunity to consider the significance of Ramsey theory. In July 1976 the Viking Orbiter 1, in the course of its search for suitable landing places, took a photograph of the Martian Cydonian plains. One of the hills photographed by the orbiter bore a striking resemblance to a face, either human or simian.

NASA published the shot of the Cydonian face because it thought

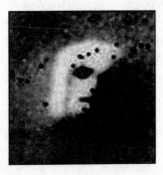

Frank Ramsey on Mars?

it was a little more attractive and amusing than most of Viking's pictures of the rutted Martian surface. The NASA scientists were unprepared for the furor that followed. Many people insisted that the face must have been artificially created, put on Mars as a message to be discovered by earthlings. A miniature industry sprang up to interpret the face and other features discovered by sharp-eyed searchers with keen imaginations. If they had had a better understanding, they would have known that the face photographed by the Viking probe was the face of Frank Ramsey—or at least of his theorem.

Ramsey's theorem can do more than explain the delusions of the gullible. In Szekeres's imagination it is only a small leap from studying points strewn across a Euclidean plane to such cosmic questions as the origin of life. "The genetic code gives you some instructions, the equivalent of saying 'the points are in the plane,' and suddenly a leaf appears on a tree" as certainly as a convex polygon emerges on the plane. Szekeres likes to tell his students that in few logical steps "you move from such a childish problem to the greatest mysteries of our existence. The whole of life in a way. It touches a little bit on this. Really, I try to persuade them that they should never listen if anyone tells you that these sorts of problems are just sort of useless flights."

Erdős never needed any convincing. He was at once taken by Szekeres's proof and, inspired, quickly found an ingenious proof of his own, one that did not require the use of Ramsey's theorem. Erdős's proof had the virtue of providing a much more accurate estimate of the number of points needed to ensure the existence of a convex polygon with any given number of sides. As Esther had observed, it takes five points to guarantee a convex quadrilateral. One of their friends, E. Makai, proved that when any nine points are drawn on a plane, five of them will inevitably form a convex pentagon. Someone noticed that 5, the solution to Esther Klein's puzzle, is equal to 2×2 plus 1, that is, $2^2 + 1$, and Makai's 9 is equal to $2 \times 2 \times 2 + 1$, or $2^3 + 1$. Combined with a bit of mathematical intuition, this led Erdős, Szekeres, and Klein (nobody remembers who thought of it first) to conjecture that the number of points needed to guarantee a convex hexagon (six sides) was $2 \times 2 \times 2 \times 2 + 1 = 2^4 + 1 = 17$. Nobody has yet succeeded in proving that seventeen points will suffice. Nevertheless, based on just two data points, the young mathematicians boldly stated the generalization that a convex polygon of n sides would always arise if $2^{(n-2)} + 1$ or more points were sprinkled across the plane. Nobody has yet been able to prove that conjecture, though Szekeres wrote that they "firmly believe [it] is the correct value."

Having triumphantly solved Esther's problem, Szekeres recalls, "there was no longer any question of me and Esther having a close relationship." Erdős, he recalls, "had some sort of emotional link to this whole setup and he then threw himself on the Ramsey theorem, and he became its greatest expert and proponent." Ramsey theory—the term is Erdős's—became an independent field of mathematics.

Erdős's paper with Szekeres would be one of his earliest and brightest gems, and he would remain enamored of Ramsey Theory and combinatorial geometry for the rest of his life. But his affection for the problem solved in the paper would always be entwined in his mind with the affection in which he held the poser of the problem and its first solver. George Szekeres and Esther Klein were engaged a year later. Erdős, almost as pleased with this result as with the mathematics contained in the paper, would always call Esther's puzzle "the Happy End Problem." George and Esther married in 1936. "I remember the wedding day," Erdős would say, demonstrating how his mind linked all events to mathematics. "It was just a day after I learned that Vinogradov had proved the odd Goldbach conjecture."

CHAPTER FIVE

ERDŐS AND THE FATE OF
WESTERN CIVILIZATION

THE probability of an academic future for a young Jewish mathematician in Hungary, even one as brilliant as Paul Erdős, was smaller than epsilon. George Szekeres's parents, aware of that, insisted that their son study chemical engineering so that someday he might run the family leather factory. Szekeres took their advice and did mathematics as a romantic sideline. Andrew Vazsonyi considered a similar course, but he made the mistake of telling Erdős, who was horrified. "I'll hide, and when you enter the gate of the Technical University, I will shoot you," he threatened. "This settled the issue," Vazsonyi said.

Erdős's violent encouragement of his friend's mathematical career did not reflect naïve optimism about Hungary and Europe generally. On their excursions into the Buda hills, when not discussing mathematics, Erdős and his circle analyzed the deteriorating political situation in Hungary. Increasingly, Erdős's Jewish friends, many of whom were active in leftist politics, found themselves "studying Jordan's theorem" inside prison walls. They were intimidated in the streets, ejected from the university, and watched by the police. "Since 1925 it was clear to my parents, and me as well, that I would have to go abroad," Erdős would recall. And as he completed his doctoral dissertation, Erdős began plotting his departure.

Initially, Erdős's parents had hoped that he might continue his studies in Germany, but as they witnessed the horrifying rise of Nazism they realized that would be impossible. At his thesis adviser's suggestion, Erdős wrote to the British number theorist Louis Joel Mordell and asked for his help in obtaining a scholarship. Erdős included a copy of his paper that contained his simple proof of Schur's conjecture about abundant numbers, which was all he needed to secure a fellowship at the University of Manchester. The fellowship, which was financed by the Royal Society, amounted to £100.

Erdős was twenty-one when he received his PhD from Pázmány University in 1934, making him one of the youngest people until then ever to receive that degree. In September he boarded a train and for the first time left Hungary. "He didn't know how to cope with the meals on the train or anything," said Anne Davenport, who, along with her husband Harold, a Cambridge University mathematician, would become a close friend of Erdős.

If Erdős was a bit overwhelmed by the mechanics of travel, he did not let it interfere with his desire to meet as many mathematicians as possible. On his way to Manchester he stopped in Zurich to visit with George Pólya, one of the authors of a famous book of problems that had occupied Erdős and his friends during their sessions at the Statue of Anonymous.

On October 1, 1934—his memory would effortlessly supply the precise date more than fifty years later—Erdős's train pulled into the station in Cambridge for another quick visit before he was to move on to Manchester. Even on his first journey Erdős managed to squeeze in as many mathematical centers as possible. He was met at the station by Harold Davenport and a young German mathematician named Richard Rado, who would become one of Erdős's most important collaborators. Rado had been one of Schur's best students but as a Jew was forced to flee Germany when Hitler came to power. Erdős and Rado had been corresponding for more than a year on mathematical issues—Erdős had sent Rado a conjecture concerning an infinite version of Ramsey's theorem that Rado refuted by return mail—so when the three met at the station they "immediately went to Trinity College and had our first long mathematical discussion," as Rado later recalled. It was in the Trinity dining hall that Erdős discovered he had never buttered bread for himself.

While at Cambridge Erdős challenged the mathematicians whom he met with an amusing conjecture of his that was not of any great mathematical importance and that, in fact, proved to be wrong. But, like many of Erdős's other casual conjectures, it would change the lives of those who worked on it. One of them, Cedric Smith, would later remark with some—only slightly strained—justification that, much as the flapping of a butterfly's wing in Montana might have caused a monsoon in India, Erdős's little conjecture might have altered the fate of Western civilization.

The fateful conjecture concerned a kind of geometrical puzzle known as a dissection. A dissection is simply the cutting up of one shape into a number of smaller shapes, as in a jigsaw puzzle or an M. C. Escher print. People have come up with all sorts of clever dissections in which, for example, the pieces that make one shape can be reassembled into another shape, a triangle cut up in such a way that the pieces can be reassembled to form a pentagon, for instance. Erdős's dissection problem, at least on the surface, seems to be much simpler than that. It's easy to dissect a square into smaller squares: a chessboard, for example, is a dissection of a square into sixty-four smaller squares. But what if none of the smaller pieces is allowed to be the same size? The result would be a design worthy of Mondrian. Erdős guessed that such a dissection was impossible. In any dissection of a square into a finite number of smaller squares, at least two must be of equal size.

Where does a guess like this come from? Why would someone have any intuition at all about how a square may or may not be dissected? Usually such intuitions are based on long hours of doodling and trial-and-error. Although mathematical proofs are based on pure logic, mathematics itself is, to a large extent, an observational science. In this case Erdős might have been influenced by first looking at the problem in three dimensions. Surprisingly, it is fairly simple to prove that you cannot dissect a cube into a finite number of smaller cubes, no two of which are the same size.

Assume for a moment that it is possible to cut a cube into smaller, unequal cubes. Each face of the cube would thereby be divided into a number of different-size squares, the bottoms of the outer layer of cubes. Concentrate on one of these faces, and particularly on the *smallest* square on that face. It can't be a corner square, since it would be impossible to fit larger squares along the two inside edges of the corner square without their overlapping. Nor could the smallest square be located anywhere along an edge, since in that case it would be sandwiched between two larger squares. The projecting edges of the two larger squares adjacent to this smallest square would fence off a region the width of the smallest square that must be filled only by still smaller squares—which, by assumption, do not exist. Therefore, the small cube must lie somewhere toward the middle of the face, surrounded on all sides by larger squares, which are the bottoms of larger cubes. This means that the top of the small cube resembles a cloistered courtyard, walled in on all sides by the faces of its larger neighbors. The top of the small cube must then, like the face of the original, be covered with even smaller cubes. The argument repeats: The smallest of these cubes must be covered by yet smaller cubes, and so on, like Jonathan Swift's fleas:

> So, naturalists observe, a flea
> Hath smaller fleas that on him prey;
> And these have smaller still to bite 'em
> And so proceed *ad infinitum*.

Therefore, the dissection of a cube into a *finite* number of unequal cubes is impossible.

Even though this argument cannot be modified to work in two dimensions, Erdős's geometrical instinct told him that a similar dissection of a square was also impossible. A Cambridge lecturer named W. R. Dean heard about Erdős's conjecture and mentioned it to some high school students to whom he sometimes gave talks about mathematics. One of the students, Arthur Stone, who would later become a friend and collaborator of Erdős, became intrigued by the question—though not until years later did he learn that the question had originally come from Erdős.

Stone won a scholarship to attend Trinity College, Cambridge, where he became a good friend of two fellow math majors, Cedric Smith and R. Leonard Brooks, and of William Tutte, who was studying chemistry. "While we knew that we were real mathematicians," Smith later wrote, "we were still broad-minded enough to talk to someone who was only a chemist." Besides, he was a good chess player. Tutte soon established his mathematical credentials among his new friends by stumping them with an ingenious mathematical puzzle.

Stone retaliated by telling his friends about Erdős's problem of dividing a square into unequal squares. Stone did not have a solution, which only made the problem more attractive. The four of them would work on the problem for the next three years.

Their first triumph was to discover that it was possible to divide a rectangle into unequal squares, though they later learned that they had been scooped by a mathematician named Moroń in 1925. They developed a surprising method of finding such rectangles that reduced the problem to analyzing an electrical circuit, and found several different rectangles, but no squares.

They remained stalled at that point for a long time, but the rectangles they found were attractive. Brooks cut up one of the rectangles into its component squares, making a jigsaw puzzle that he gave his mother to solve. She succeeded in assembling the pieces into a rectangle, but to everyone's surprise it was not the same rectangle Brooks had started with. It was as if he had given his mother the pieces to a puzzle of the Empire State Building, which she assembled to form a picture of the Brooklyn Bridge. Something strange was going on, but what?

Bill Tutte thought about it and finally came up with an explanation. Brooks's mother's discovery revealed a symmetry in the underlying equation that before long helped the students to find a way to dissect a square into unequal squares. Erdős was wrong. Unfortunately, they were once again scooped, this time by less than a year by a mathema-

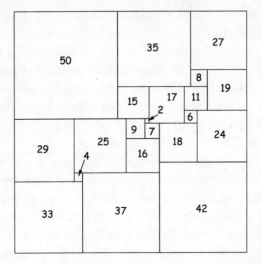

The smallest square that can be dissected into unequal squares.

tician named R. P. Sprague. Squared squares, as these objects are known, have been exhaustively studied in the intervening years. In 1978 A. J. W. Duijvestin discovered that the smallest number of unequal squares into which a square can be divided is twenty-one. His dissection is pictured here.

In 1939 the employment outlook for academics in Britain was grim. Cedric Smith remembers bumping into one of Tutte's tutors, Patrick Duff, who asked him, "Do you know Tutte?"

"Yes," Smith replied.

"We're very worried. He's no good."

"He's got a first-class degree."

"His supervisor is disappointed."

"Well, he's very good at maths."

"Prove it," Duff demanded.

Smith sent the College officials a letter containing what he hoped would be considered proof. Foremost among the evidence was the work Tutte had done disproving Erdős's conjecture. If the Trinity elders were impressed, they never let on. But when World War II broke out, Tutte was recruited to join a secret program at Bletchley

Park that employed some of Britain's best mathematicians, including Alan Turing. Apparently Smith's letter had been noticed.

The Germans relied on the security of elaborate codes, the Ultra and the Triton, which were generated by the so-called Enigma Machine. The Poles had examined an Enigma machine as it was transported across the Polish Corridor, and their reconstruction of the device provided the Bletchley Park mathematicians with a good understanding of the code. Breaking the code, which was constantly shifting, proved to be extremely difficult. "Very plausible rumor," Smith writes, "says that Tutte supplied the vital clue." According to the historian Paul Johnson, cracking the "Ultra played a part as early as 1940 by helping to win the Battle of Britain. More important, the breaking of the German 'Triton' code by Bletchley in March 1943 clinched the Battle of the Atlantic." The decrypted broadcasts of German U-boat captains allowed the Allies to intercept and destroy vital German supply ships. If British mathematicians did not actually win the war, they shortened it. "Civilization was saved. And it all began with a conjecture by Erdős," Smith concludes. If Erdős's conjecture did not directly, or even indirectly, save Western civilization, it did what so many of his conjectures would do; it engaged a mind and changed a life.

After his brief visit to Cambridge, Erdös moved on in October 1934 to the University of Manchester to take up his work with Louis Joel Mordell. In the early years of the century, Mordell, a math-obsessed teenager attending Philadelphia's Central High School, would pore over used mathematics books he purchased for a nickel or a dime from a local bookstore. Some of those books contained intriguing examples from Cambridge University tripos, or honors, examinations, which inspired Mordell with the determination to attend Cambridge. After racing through high school in two years, Mordell scraped up the money for passage to England, where he placed first on the Cambridge scholarship exam. Mordell went on to a distinguished career in number theory, a subject in which he was largely self-taught, since few British mathematicians at the time were drawn to it. His most famous achievement was the "Mordell Conjecture," which, when finally proved in 1984 by Gerd Faltings, ultimately led to Andrew Wiles's proof of Fermat's conjecture.

The ten-year-old Erdős
with his parents.

(Courtesy of Janos Pach)

Paul Erdős with his mother
at Lake Balaton, 1916–17.

*(Reproduced with permission from the Center for Excellence
in Mathematical Education, Colorado Springs)*

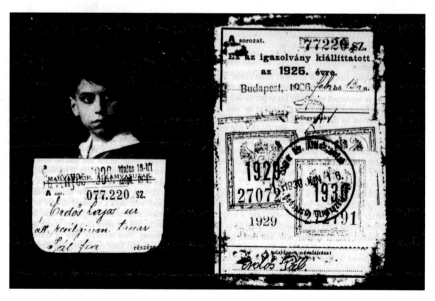

Erdős's half-fare ticket to ride on the Hungarian State Railway.

(Courtesy of Janos Pach)

Erdős with the number theory group at the University of Manchester in 1937 or 1938. Erdős can be identified as the only one not wearing a necktie. Chao Ko stands at far right, next to Guy Davenport. Louis Mordell, who invited Erdős to come to Manchester, is the second figure to the left of Erdős.

(Reproduced with permission from the Center for Excellence in Mathematical Education, Colorado Springs)

Erdős at the Institute
for Advanced Study
in 1941. When Erdős's
fellowship at the
Institute was not
renewed, he began his
life of wandering.

(Courtesy of Janos Pach)

Mathematicians can work anywhere. Here, Erdős works
on a problem in one of his favorite settings, the mountains of Hungary.

*(Reproduced with permission from the Center for Excellence in
Mathematical Education, Colorado Springs)*

Anyuka: Erdős's mother, Anna, at the guest house of the Hungarian Academy of Sciences in Matrahaza, where "she was very much the Queen Mother" of mathematics.

(Courtesy of Janos Pach)

Erdős and Richard Rado, sometime in the early 1950s.

Erdős engaged
in a favorite activity,
talking to elementary
schoolchildren—
"epsilons"—about
mathematics in
Sztalinvaros,
Hungary, in 1955.

(Courtesy of Janos Pach)

Erdős and Alfred Renyi
in Århus, Denmark,
in 1957.

(Courtesy of Janos Pach)

Erdős, Szekeres, and Turán take time out in 1958 to play a game of ping-pong . Erdős was a deceptively tough opponent: "One could not help thinking that in his nervous system impulses somehow travelled a lot faster."

(Courtesy of Janos Pach)

Bela Bollobás met Erdős when he was fourteen, and they wrote their first paper together when he was seventeen. Bollobás would become one of Erdős's leading collaborators. "He always knew which problems were good for whom. . . . For different horses you need different courses."

(From the film N Is a Number: A Portrait of Paul Erdős, *by George Paul Csicsery)*

Erdős delivered his "Sixty Years of Mathematics" lecture at Cambridge University in June 1991, the day before being honored with a prestigious honorary doctorate.

(From the film N Is a Number: A Portrait of Paul Erdős, *by George Paul Csicsery)*

The Thinker: Erdős at work in Colorado in 1992.

(Reproduced with permission from the Center for Excellence in Mathematical Education, Colorado Springs; photograph by Alexander Soifer, 1992)

Erdős could never resist an "epsilon," whether friend or stranger. Here he is dandling Isabelle Soifer, the daughter of one of his collaborators.

(Reproduced with permission from the Center for Excellence in Mathematical Education, Colorado Springs; photograph by Alexander Soifer, 1991)

Portrait of the mathematician.

(From the film N Is a Number: A Portrait of Paul Erdős, *by George Paul Csicsery)*

Manchester already had a good reputation as a mathematical center in 1922 when Mordell took over as head of the mathematics department, and under his guidance it grew into a leading center. Mordell attracted many foreign visitors to Manchester. Like Erdős, they were escaping the rising tyrannies of their native countries. Rado and Davenport would also wind up in Manchester during Erdős's four-year stay. Most of the papers Erdős wrote in these years were concerned with number theory, which reflected the taste of the group of mathematicians assembled by Mordell as well as Erdős's natural interest. In fact, of Erdős's first sixty papers—far more than most mathematicians write in a lifetime—all but two were about number theory.

The dominance of number theory among Erdős's early papers reflects his love of the field Gauss called "the Queen of Mathematics," but it is also an indication of the role that fashion plays in even the rational domain of mathematics. Erdős's interest in graph theory dated back to his class with Dénes König. His love of combinatorics, graph theory's close kin, was fueled by his work on the Happy End problem. Graph theory was derided by most mathematicians as "the slums of topology," and combinatorics had been relegated to an even less savory neighborhood. Visiting those neighborhoods was fine for an occasional lark, but no respectable mathematician would want to live there.

But even in papers about number theory, Erdős managed to smuggle in some graph theory. In 1938 Erdős wrote a paper called "On sequences of integers no one of which divides the product of two others and on some related problems." The title says it all, or at least as much as can be said here. The remarkable accomplishment of this paper was that in it Erdős reduced what was a problem in pure number theory to a problem of finding graphs having certain properties, thus joining together two fields that seemed to have little in common, and in the process adding strength to the one while endowing the other with respectability. This paper is also noteworthy because it marks one of the rare instances when Erdős failed to appreciate the true significance of his own work.

Erdős failed to anticipate that the graph theory problem he had solved in this 1938 paper on integer sequences was the prototype of what would become the important new field of extremal graph theory. The simplest question of extremal graph theory is this: Consider a graph with n vertices (points). What is the maximum number of edges (lines) that this graph can have without containing a triangle? For a graph with four vertices the answer is clearly 4: The four edges form a rectangle; a fifth edge would be a diagonal of the rectangles, which divides it into two triangles (see diagram on this page).

The general answer is that triangles can be avoided if the number of edges is at most one-fourth the number of vertices squared. It is not *extremely* difficult to prove this result (though beyond the scope of this book), but the proofs quickly get much more difficult when shapes other than triangles are to be avoided. Within a couple of years Paul Turán would come to understand that these extremal problems—problems concerning the extremes under which structure emerges—were so interesting, beautiful, and rich they deserved study in their own right. He wrote a paper that would form the basis of the new discipline of extremal graph theory. Erdős became one of the leading contributors to extremal graph theory, which is filled with his theorems and conjectures. Still, Erdős would come to chastise himself because he always felt that he "should have invented" the field that he came to dominate.

Erdős would compare his "blunder" with that of Sir William Crookes, the physicist who invented the cathode ray tube. Crookes noticed that his tube fogged photographic film sealed inside a light-tight envelope. From that observation, Erdős remarked, Crookes

In a graph on four vertices and five edges a triangle is inevitable.

concluded that "nobody should leave film near the cathode-ray tube." But when Wilhelm Roentgen ruined some film in the same way a few years later, the experience led him to invent the x-ray machine. The moral, according to Erdős? "It is not enough to be in the right place at the right time. You should also have an open mind at the right time."

Erdős's other foray into combinatorics while he was at Manchester, a paper written with Rado and a young Chinese mathematician named Chao Ko, also contained intimations of extremal graph theory. The paper, which included an important result known as the Erdős-Ko-Rado theorem, was completed in 1938. But, in part because of the mathematical community's lack of interest in combinatorics, the paper was not published until 1961, when it became an "instant classic."

Having become accustomed to riding trains and finding them a good place to work, Erdős made frequent visits to Cambridge and other mathematical centers. "His *Wanderlust* was already in evidence," his disciple Béla Bollobás wrote in a biographical essay. "From 1934 he hardly ever slept in the same bed for seven consecutive nights, frequently leaving Manchester for Cambridge, London, Bristol, and other universities."

Throughout his Manchester years Erdős maintained close contact with his friends back in Hungary and visited home at least three times a year, at Christmas, Easter, and the summer vacations. The political situation in Budapest, as in the rest of Europe, was becoming steadily worse. By 1936 Erdős's friend László Alpár had been imprisoned ten times. "A great number of people advised me to go to Paris to continue my education there," Alpár recalled. "In Paris the Popular Front was in power and uncle Erdős [Erdős's father] helped me raise the money to cover my travel expenses and to be able to financially survive the first period of my stay there." Vazsonyi would also emigrate to Paris, and Esther and George Szekeres would later escape to Shanghai.

As the reports on the radio grew grimmer, Erdős began to plot his escape from Europe. In 1937 he applied for a fellowship to the Princeton Institute for Advanced Study, only to be informed by his countryman and Institute member John von Neumann that although "we are ... very desirous of making a visit from you possible," no money was available at the time. Within a few months the budget for the following academic year was approved and Erdős was offered a stipend of $1,500 to visit the Institute in 1938–39.

In March 1938 Hitler's troops rolled into Austria, which the Führer declared to be part of the Third Reich. That summer Erdős visited Esther and George Szekeres, who were vacationing in the picturesque Bükk Mountains. The friends believed that they might never meet again, as Szekeres poignantly recalled almost sixty years later. "Those were already grim times, dark clouds were gathering over Europe and Erdős had no doubts that, if there was any chance, we should leave the country (which we did a year later). He saw the future more clearly than Chamberlain, Britain's Prime Minister. The moment of farewell arrived; Eszti [the diminutive of Eszter, the Hungarian form of Esther] and I watched with a lump in our throats as the bus disappeared with him in the dust of the road." To underscore the reality of those fears, Szekeres recounts meeting Géza Grünwald, who was also rambling in the Bükk of Hungary, a few days later. Grünwald was another of their close friends and one of Erdős's earliest collaborators. "A few years later Géza perished in a forced labor camp on the Eastern front, not fighting the 'enemy' but at the hands of his own compatriots."

On September 3 Erdős became alarmed by Hitler's demands to annex the Sudetenland, a German-speaking region of Czechoslovakia. After saying his fast farewells to his Budapest friends—many of whom would perish in the war—and his beloved *Anyuka* and *Apuka*, Erdős boarded the train that carried him on an indirect route through Italy to Zurich, Paris, and finally London. On September 28, Erdős embarked on the *Queen Mary* bound for New York, passing through Ellis Island on October 3 on his way to Princeton. He would not see Budapest again for ten years.

CHAPTER SIX

PARADISE LOST

T H E Institute for Advanced Study had been operating for exactly five years when Paul Erdős arrived on October 4, 1938. Located on the outskirts of the town of Princeton, New Jersey, on a square mile of wooded land, isolated from the cares of the world, the Institute was, in the words of its creator, the American educational reformer Abraham Flexner, truly "a paradise" of the intellect.

Eight years earlier Flexner had been approached by the would-be philanthropists Louis Bamberger and his sister, Caroline Bamberger Fuld. The siblings had until recently been the owners of Bamberger's, a department store that was the fourth largest retail chain in the nation. Exhibiting the same insight and luck that had helped them build their empire, the Bambergers sold out to the R. H. Macy Company just six weeks before Black Monday, 1929, the day the stock market crashed. The Bambergers were as generous as they were lucky. They were determined to use their fortune to benefit the people who had helped them to amass it, the shoppers of New Jersey.

Since their first impulse was to endow a medical school, they paid a visit to Flexner, a distinguished doctor and educator famed for exposing the corrupt and fraudulent medical schools of the day.

Flexner had his own ideas about how best to spend the Bamberger fortune. Flexner painted for his visitors his vision of a research institution unlike any the world had seen since perhaps the time of Pythagoras. Like the Pythagorean school, the Institute that Flexner conceived would be "a haven where scholars and scientists could regard the world and its phenomena as their laboratory, without being carried off into the maelstrom of the immediate." America had lots of medical schools, but there was nothing anywhere quite like the Institute of Flexner's imagination. The Bambergers were enthralled and were soon writing big checks to subsidize Flexner's dream.

In Europe the "maelstrom of the immediate," fueled by fascism, was spinning out of control, and many of the world's most brilliant scientists and mathematicians were desperate for a safe haven. Chief among them was the most famous scientist in the world, Albert Einstein. Flexner's first coup was to sign Einstein as the Institute's first professor, which immediately propelled it to international fame. With the cachet of an Einstein, not to mention opulent salaries and idyllic working conditions—there were no teaching duties because there were no students, only professors and postdoctoral "workers"—the Institute was soon one of the prized destinations for the world's intellectual elite, at least mathematicians and theoretical physicists, whose work was of an empyrean purity that few disciplines could match. When a visitor to the Institute asked Einstein to see his laboratory, the great physicist withdrew a fountain pen from his breast pocket with a flourish and announced, "Here!" That's the way they liked it at the Institute for Advanced Study.

To Erdős the Institute must have seemed every bit the paradise that Flexner had promised it would be. J. Robert Oppenheimer, who would later become the Institute's director, called it an "intellectual hotel"; to Paul Halmos, another Hungarian mathematician who was at the Institute with Erdős, it was nothing less than a "country club for math." "How did anyone—how did I—ever get any work done at Princeton?" Halmos would wonder.

Between long walks, loafing in the common room, and endless games of Go, it was hard to imagine when work got done. That Erdős and other mathematicians became addicted to Go at the Institute is easy to understand. The ancient Asian game is deceptively simple, played by alternately placing black and white stones (at the Institute they used thumb tacks) at the intersections of a 19x19 rectangular grid. A game of Go, viewed from the right perspective, is really nothing more than a problem in graph theory. If, as G. H. Hardy wrote, "chess problems are the hymn-tunes of mathematics," a game of Go is a cantata. IBM's chess-playing supercomputer Deeper Blue beat the world champion Gary Kasparov in a chess match, but the best Go-playing computer cannot beat even a good amateur at Go, and is unlikely to do so any time soon. That's why Go was a perfect leisure activity for the brainy Institute crowd.

Despite the fun and games, Halmos, Erdős, and the others managed to get an amazing amount of work done. In a lifetime of legendary productivity, the year and a half Erdős spent at the Institute stands out. Everywhere he looked he found problems to solve and colleagues with whom to collaborate. Erdős astonished even the brilliant Institute crowd with his natural mathematical ability. Once, in the common room of Fine Hall, he overheard two mathematicians discussing a problem in dimension theory, which was a part of topology about which Erdős knew almost nothing. They were struggling with the unsolved problem of determining the dimension of the set of rational points in Hilbert space. Never mind what that means; Erdős did not understand the question either. Smart money was saying that the answer was either zero or infinity, but the mathematicians at the blackboard—Witold Hurewicz and Henry Wallman, two of the world's leading experts in the field—weren't making any headway.

"What is the problem?" Erdős asked. Impatient at being interrupted, they hastily told him.

"What is dimension?" he then asked, revealing his hopeless ignorance. He wasn't clear on what a Hilbert space was either. To silence the intruder, Wallman gave him a hasty definition of dimension. To their relief, Erdős went away. One hour later he returned with the solution to their problem. To the experts' surprise, the answer was 1.

Erdős's paper, Halmos wrote, came out a year later, an "important contribution to a subject that a few months earlier Erdős knew nothing about."

Erdős would always describe the year he spent in Princeton as the most successful of his career. He continued to mine the integers for strange and unexpected results. In one solo effort, for example, Erdős proved that the product of any number of consecutive integers is never a perfect square. Results like that, so concise and unambiguous, give a reassuring sense of the orderly structures of arithmetic. But that year Erdős would also show, in a paper that would mark one of his greatest achievements, that beneath the apparent regularities of the integers, chaos lurked. While Albert Einstein, the Institute's most famous resident, was trying to disprove quantum theory and thus prove that God does not play dice with the universe, Erdős and a young Polish mathematician named Mark Kac were demonstrating that the Supreme Fascist was having fun with the integers.

Kac had recently arrived from Poland to work at Johns Hopkins on the theory of probability with Aurel Wintner, a Hungarian mathematician who had emigrated to the United States in 1930, though the mathematics was not the only reason for Kac's move. In a joke that was then popular in Europe, one man asks another, "Are you Aryan, or are you taking English lessons?" Kac was not Aryan, so he hastily and somewhat haphazardly learned a bit of English before coming to Baltimore. His mathematical vocabulary was good, but he had trouble ordering food; in a restaurant he'd rarely get a dish that even vaguely resembled what he had asked for. Kac solved the problem of lunch by sticking to a local drugstore and learning to say, "Cream cheese sandwich and coffee" intelligibly. The counterman would invariably respond, "On toast?" That being beyond Kac's linguistic competence, he would smile inanely, which seemed to work. Kac looked up "toast" in his pocket dictionary and found this definition: "Gentlemen, the King!" "Having been logically conditioned, I assumed that 'on toast' must be some kind of salutation and I proceeded on this assumption," Kac wrote. Now, for two weeks, every time the waiter asked, "On toast?" Kac would bow politely and reply, "On toast!" After a while Kac, sensing something was wrong, asked a friend about this. After he finished laughing, his friend then asked: " 'Why didn't you at least once answer in the negative? You would have soon known what "on toast" meant.' 'I didn't want to risk being impolite,' Kac replied, and he laughed again."

While still in Poland, studying with the great Polish mathematician Hugo Steinhaus, Kac had become bewitched by probability theory, more particularly by the normal distribution, better known to generations of students as the bell curve. The familiar hump of the bell curve occurs wherever randomness is at work. Measure the heights of a group of twelve-year-olds: most will cluster around some average value, with the number of taller and shorter children declining as one departs farther from this average. The same shape describes the distribution of IQs—most of the population lies in the central hump around 100, with Erdős and the crowd at the Institute somewhere in the narrow shoulder on the right. The same is true of life spans or coin tosses. The marble tread of an ancient staircase, worn by the innumerable footsteps of generations, takes on the shape of an inverted bell curve; the wear is deepest in the center, where most people step, and gently tapers off to either side, a physical embodiment of the underlying mathematical principle.

The normal distribution was discovered in 1733 by the French mathematician Abraham De Moivre. When De Moivre was eighteen, in 1685, Louis XIV revoked the Edict of Nantes, which had guaranteed civil rights to Protestants in predominantly Catholic France. De Moivre, a Protestant, was imprisoned for his beliefs. When he was released two years later, De Moivre fled France for England, which had outlawed Catholicism following the "Glorious Revolution" of 1688. Settling in London, De Moivre was unable to secure an academic position. He supported himself largely by tutoring mathematics. Most afternoons, after his tutoring was done, he could be found at Slaughter's Coffee House in St. Martin's Lane, where he sold his expertise on matters of probability to gamblers who threw dice, played cards, or sold insurance.

De Moivre became interested in a problem that is dear to the heart of gamblers: How can you tell if a coin is fair? A fair coin is as likely to land heads up as tails. In a hundred tosses a fair coin should come up heads fifty times. But that is only an average; sometimes forty-three heads will come up, sometimes sixty-two. In fact, any outcome between 0 and 100 heads is possible, but some are very unlikely. The key to determining whether a coin is fair is knowing the probability of the various outcomes. In Tom Stoppard's play *Rosenkrantz and Guildenstern Are Dead,* the two title characters wager on the toss of a coin that may or may not be fair. Guildenstern, betting on tails, is understandably upset when the coin lands heads up eighty-five times in a row. He confronts Rosenkrantz, who sees nothing wrong.

> GUIL: No questions? Not even a pause?
> ROS: You spun them yourself.
> GUIL: Not a flicker of doubt?
> ROS: Well, I won—didn't I?
> GUIL: And if you'd lost? If they'd come down against you, eighty-five times, one after another, just like that?
> ROS: Eighty-five in a row. *Tails?*
> GUIL: Yes! What would you think?
> ROS: Well. . . . Well, I'd have a good look at your coins for a start!

Mathematics helps where self-interest enters the picture. Guildenstern says, "The equanimity of your average tosser of coins depends upon a law, or rather a tendency, or let us say a probability, or at any rate a mathematically calculable chance, which ensures that he will not upset himself by losing too much nor upset his opponent by winning too often." Guildenstern's suspicion that the coin was tweaked could have been confirmed by a knowledge of De Moivre's normal distribution, which gives the probability distribution of outcomes to be expected when a coin is tossed many times.

With the help of some calculus, recently invented by his friend Isaac Newton, and some counting tricks of Blaise Pascal, De Moivre found the probability of every possible outcome of a long series of coin tosses. As expected, the formula describes a curve that peaks at equal heads and tails and falls off symmetrically in both directions. On either side of the central peak the normal curve is shaped like a good sledding hill, descending steeply at first and then flattening out. The probability that the number of heads is between, say, 30 and 60 is simply the area under the curve between the two limits.

It is clear from examining the shape of the normal distribution that the outcomes of several series of coin tosses will cluster in a narrow region around the central peak, which marks the expected average. In the case of a coin tossed a hundred times, the average expected result, or mean, is fifty heads. But variations from the mean are expected, and to quantify these Moivre invented an important measure of the expected range of these variations called the standard deviation. In a normal distribution about two-thirds—68 percent—of all observations fall within one standard deviation of the mean, and 95 percent of all observations fall within two standard deviations. For a fair coin the standard deviation is the half square root of the number of tosses. The standard deviation of a coin tossed a hundred times is, therefore, 5, or half the square root of 100. Toss a coin 100 times and two-thirds of the time the number of heads will lie between 45 and 65; 95 percent of the time the number of heads will lie between 40 and 60.

Darwin's nephew, Francis Galton, aptly described the normal distribution as "the supreme law of Unreason." Indeed, after almost three centuries the theory of probability can still defy the reason of even the greatest mathematicians. On Erdős's last visit with his old friend Andrew Vazsonyi in Santa Rosa, California, Vazsonyi "for some unknown reason" decided to test Erdős's probabilistic intuition with a popular puzzle known as the Monty Hall problem. By that time Erdős was one of the world's leading authorities on probability. One of his greatest achievements, the "probabilistic method," is often simply called the

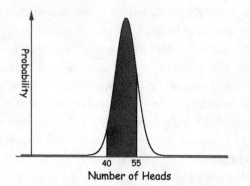

The probability that a coin tossed 100 times will come up heads between 40 and 55 times is given by the area under the bell-shaped normal distribution curve.

"Erdős method," so in a sense the name Erdős had become synonymous with probability. Vazsonyi expected that Erdős would immediately penetrate to the heart of the Monty Hall problem as he had with many far more difficult problems. Vazsonyi was mistaken.

The Monty Hall problem had been circulating throughout the mathematical community for several years when Marilyn vos Savant—the self-styled world's smartest woman—published it in her weekly column in *Parade* magazine in September 1991. The problem is a variant of the kind of good-natured cons that Monty Hall, the host of the game show "Let's Make a Deal," was fond of running.

Imagine you are a guest on the show and Monty allows you to choose from one of three doors. Behind one door is a fabulous prize—a million dollars in gold bullion, a trip to the moon on gossamer wings, or whatever your heart desires. Behind the other two doors are joke prizes of no worth—a goat, say. Before Monty opens the door you chose, he opens another door and reveals a goat. Because two doors conceal goats, Monty can still show a goat, no matter which door you chose. Monty then makes an offer: You may stick with the door you originally chose or switch to the other unopened door. What do you do?

Vazsonyi told Erdős that the correct strategy was to switch, "and I fully expected to move on to the next subject. But Erdős, to my surprise, said, no, that is impossible, it should make no difference." Erdős had fallen into the trap that had captured the many mathematicians who fired off vitriolic letters to vos Savant that they would later regret. One, who has suffered enough embarrassment and will therefore remain anonymous here, wrote: "You blew it! Let me explain: If one door is shown to be a loser, that information changes the probability of either remaining choice—neither of which has any reason to be more likely—to ½. As a professional mathematician, I'm very concerned with the general public's lack of mathematical skills. Please help by confessing your error and, in the future, being more careful." Other mathematicians were not as polite.

The professor's justifiable concern for the public's mathematical skills nonwithstanding, switching doors is by far the best course of action. Perhaps the easiest way to understand the solution is to notice that your probability of initially having chosen the correct door was one in three. This probability never changes, even after Monty shows you the goat. Since the odds of the prize being behind your door are still one in three, the probability that it is behind the other door must be two in three.* Switching will therefore double your chances of winning the prize. Erdős and all the irate mathematicians who wrote to *Parade* had failed to understand that by showing the goat Monty was providing valuable information.

Vazsonyi explained all of this, using the language of mathematics, but Erdős remained unconvinced and, like vos Savant's correspondents, began to get upset. "At this point I was sorry I brought up the problem," Vazsonyi recalls. Eventually, frustrated by Vazsonyi's explanations, Erdős stormed away. When Erdős returned an hour later he shouted at Vazsonyi, "You are not telling me why to switch! What is the matter with you?" Erdős became convinced of the wisdom of that course of action only after Vazsonyi showed him a simulation on his computer,† but he still was frustrated by his inability to understand intuitively why switching worked. Erdős was finally mollified a few days later after his close friend Ron Graham, a mathematician at Bell Labs, explained the problem to his satisfaction.

If even great mathematicians like Erdős can be fooled by the workings of the laws of chance, what hope was there for the hapless Guildenstern? Underlying the behavior of a fair coin and the solution to the Monty Hall problem is a concept known as statistical independence. Like many seemingly common English words, the word "independence" has a precise mathematical definition that is at best only approximated by its dictionary definition. Words like "group," "field," and "information" mean different things to a mathematician and a historian. This can lead to awful puns, but it can also result in a great precision of expression. Two events are independent when the

---

* Some people find this solution easier to understand by thinking of a variation on the problem. Imagine that instead of three doors there were 1,000,000. You choose a door and then Monty opens up 999,998 other doors, revealing goats. Now it should be pretty clear that switching is by far the best thing to do. After all, your chance of having chosen the correct door was initially a million to one. Now you know that either you had been astronomically lucky in your choice or the prize is behind the one remaining unopened door.

† For those who are still not convinced and are not skilled with computers (Vazsonyi used an Excel spreadsheet), it's a good idea to simulate the Monty Hall problem with playing cards. After only a dozen trials the wisdom of switching should become obvious.

outcome of one does not affect the outcome of another. The chance of a coin landing heads up is independent of whether it has previously landed heads up once or even eighty-four consecutive times. In an optimistic moment, Guildenstern viewed his streak of bad luck as "a spectacular vindication of the principle that each individual coin spun individually is as likely to come down heads as tails and therefore should cause no surprise each individual time it does."

To find the probability that two independent random events will *both* occur, multiply together the probabilities of each one occurring *independently*. This prescription is the mathematical definition of independence. So, the probability of throwing two heads in a row is $1/2 \times 1/2 = 1/4$, since the probability that a single coin comes up heads is $1/2$. The probability of Guildenstern throwing 85 heads in a row is $1/2 \times 1/2 \times 1/2 \times \ldots$ 85 times, or $1/2^{85}$, which is about 1 in 4×10^{25}. That is another one of those numbers that is so large that it might as well be infinite. Even if he could toss his coin many trillions of times per second, Guildenstern could not reasonably expect such a run of heads before all the stars burned themselves to cinders. Either there was something fishy with that coin, or Guildenstern had managed to hit the cosmic lottery.

What Kac had started to realize in Poland was that independence also occurred in the orderly world of mathematics, and therefore the techniques of probability theory might be applicable in all sorts of places where nobody had ever suspected chance played any role. In number theory, for example, Kac noticed that the divisibility of integers could be viewed in a probabilistic light; since half of all numbers are divisible by 2, then the probability that some random number is divisible by 2 is $1/2$. Similarly, the probability that a number is divisible by 3 is $1/3$, and the probability that one is divisible by 6 is $1/6$. Kac wrote this simple observation as the arithmetical equation:

$$1/6 = (1/2) \times (1/3).$$

This says nothing more than that if a number is divisible by 6 it must be divisible by *both* 2 and 3. But writing it in that way reminded Kac of the law of multiplication of probabilities for independent events. Working with Steinhaus, Kac had learned that "Where there is independence there must be a normal law." He began to sense that some-where hidden amid the divisors of integers must be a bell curve, the signature of the normal law. But where?

One property of a number that theorists find useful and interesting to investigate is how many distinct prime factors it possesses, since that gives one measure of how nearly prime it is. Primes, which are divisible only by 1 and themselves, have only a single distinct prime divisor. The number 10, which is not prime, has two distinct prime factors, 5 and 2; 9 has only one distinct prime factor, 3; and 30 has three distinct prime factors, 2, 3, and 5. Kac's insight was to realize that just as the probability that a tossed coin will come up heads is unaffected by the outcome of previous tosses, the probability that a number is divisible by one prime is independent of whether it is divisible by any other. Since the number of heads and tails expected after a large number of coin tosses obeys a normal distribution, it seemed to Kac that the number of distinct prime factors should obey a similar law. In other words, Kac guessed that if he were to examine all numbers less than a million, say, he would find that most of them had the same number of distinct prime factors, in much the same way that when a coin is tossed a hundred times it will come up heads about fifty times. Of course, sometimes the coin will come up heads sixty times out of a hundred tosses or, rarely, only five. Deviations from the expected outcome follow the normal distribution law. Kac reasoned that, similarly, a few numbers would have many distinct prime factors and some would have only a few, and these deviations from the average would also obey a normal distribution law.

The problem was that Kac knew very little number theory and soon ran into a difficulty that he was unable to overcome. So he put aside the unfinished problem until March 1939, when he took the train from Baltimore to Princeton to give a talk. Erdős was in the audience, but when he saw that Kac was talking about probability, a subject about which he knew as little as Kac knew about number theory, he began to doze. Toward the end of his talk Kac uttered the phrase "prime divisors," and Erdős, like a lover hearing the whis-pered name of his beloved, was instantly awake and alert. Erdős stopped Kac to ask him to repeat what the difficulty was. "Within the next few minutes, even before the lecture was over, he interrupted to announce that he had the solution," Kac wrote. The proof, Erdős would later declare, was right from The Book.

The result of this serendipitous collaboration is now known as the Erdős–Kac theorem, which states the astonishing fact that the number of distinct prime factors of the integers less than N follow more or less a distribution with the same shape as the number of heads for a coin tossed N times, or, for that matter, the distribution curve of chest measurements of Scottish soldiers. In his charming memoir *Enigmas of Chance*, Kac would later write: "The reader, I hope will forgive my lack of modesty if I say that it is a beautiful theorem. It marked the entry of the normal law, hitherto the property of gamblers, statisticians and observateurs, into number theory and . . . it gave birth to a new branch of this ancient discipline."

The "new branch" would not be established for ten years. Erdős's and Kac's paper, though published in 1940, languished virtually unnoticed for more than a decade. The turmoil of World War II undoubtedly contributed to the paper's neglect, but the novelty of the result probably played an even larger role. The Erdős–Kac theorem was the result of one of the strange interdisciplinary marriages that were an Erdős specialty; the very name of the field to which it gave birth, probabilistic number theory, would have seemed an oxymoron ten years earlier. As Hardy remarked, "317 is a prime not because we think so, or because our minds are shaped in one way or another, but because it is so, because mathematical reality is built that way." As the mathematician Joel Spencer explains, "the number 219 does not have two prime factors with probability .93—it absolutely, definitely has two prime factors, 3 and 73. Yet somehow—and this is the surprise—in the aggregate the answers behave as if the SF were flipping coins." As in so many collaborations yet to come, Erdős, with his wide knowledge and open brain, functioned as a critical catalyst. Kac felt that nobody else but Erdős could have helped to bring his ideas to fruition. "It would not have been enough, certainly not in 1939, to bring a number theorist and a probabilist together. It had to be Erdős and me: Erdős because he was almost unique in his knowledge and understanding of the number theoretic method of [the Norwegian mathematician] Viggo Brun, which was the decisive end, and . . . the deepest of the ingredients, and me because I could see independence and the normal law through the eyes of [Kac's adviser Hugo] Steinhaus."

During his fellowship year at the Institute Erdős wrote another seminal paper on the basis of what would later be known as probabilistic number theory with Aurel Wintner, Kac's supervisor at Johns Hopkins. He also collaborated by mail with Turán on papers regarding the theory of interpolation, which concerns the methods for estimating the values of functions when given only a few points. It is perhaps a significant indication of his future prospects at the Institute that despite this outpouring of creativity, Erdős did not publish any joint papers with mathematicians from the Institute.

Erdős also spent long hours talking with the great logician Kurt Gödel. Both were devout Platonists, absolutely certain that the objects of mathematics—points, primes, polynomials, and all the rest—are as real as goldfish or gluons. "It seems to me," Gödel said, "that the assumption of such objects is quite as legitimate as the assumption of physical bodies and there is quite as much reason to believe in their existence." The Book, *sans* binding perhaps, was to Gödel as real as any volume in the Library of Congress, despite his having proved a few years earlier that some pages of The Book had to be blank or missing; within any axiomatic system there must be some statements that can be neither proved nor disproved. What Gödel's theorem meant was that even in mathematics, the last refuge of certainty, not all truth is knowable and not all problems can be solved. Not that any of that bothered Erdős much, because he knew there were always problems enough to keep him busy, and he had a sure instinct for which ones were solvable. He was fond of quoting Saharon Shelah, an Israeli who is one of the most brilliant problem solvers in mathematics: "I am an opportunist, I do what I can do." This attitude may explain Erdős's predilection for working in relatively new and unexplored fields like combinatorics. "If there is anything in number theory I can do, I certainly do it," he once explained. "But, you see, some of the problems in number theory are enormously difficult, and many of these classic problems are very, very hard to make progress in. Combinatorics is a much newer field, and there are many more problems that are still accessible."

Gödel's logical mind almost prevented him from becoming a citizen of the United States. When preparing for his citizenship examination, Gödel read the Constitution with more logical care than any Supreme Court justice ever had. And he found contradictions. Gödel concluded that according to the Constitution, the very axioms of democracy, the United States could legally be turned into a dictatorship. Gödel's friends cautioned him that he should keep quiet about his conclusions during his examination. Nevertheless, when the examiner observed sympathetically that while Gödel had been a citizen "under an evil dictatorship, but fortunately [such a thing is] not possible in America," Gödel jumped up to correct him. "On the contrary," he said, "I know how that can happen." Fortunately, his friends managed to constrain him until the examiner had sworn him in as a citizen.

Gödel was extremely self-critical and published infrequently. That infuriated Erdős. "He could certainly have done more things," Erdős said. "He had a proof that the axiom of choice was independent [an important problem in set theory], but he didn't like the proof." They studied the philosopher Leibniz, whose criticism of the infinitesimal quantities that Newton used to create calculus still resonates in mathematics today. As much as Erdős enjoyed his conversations with Gödel, he found them exasperating. Erdős berated Gödel, saying: "You became a mathematician so that people should study you, not that you should study Leibniz."

Not surprisingly, Erdős would always rate his countryman, dapper, urbane John von Neumann, as one of the most brilliant people he had ever met—everyone did. At six von Neumann could joke with his father in ancient Greek and could multiply and divide eight-digit numbers in his head. To demonstrate his photographic memory he would glance at a page of the Budapest telephone directory and then repeat every name, number, and address. Once a friend asked von Neumann if he had read *A Tale of Two Cities*. He had, years earlier, which he proved by reciting as much of the first chapter as his friend could bear. When still in his early twenties von Neumann had helped to put the new quantum theory of physics on a sound mathematical basis by showing that what had appeared to be two distinct competing theories—the wave equations of Erwin Schrödinger and the matrix theory of Werner Heisenberg—were actually one and the same. Von Neumann went on to invent the digital computer, the theory of games, self-reproducing automata, and volumes of path-breaking mathematics, both pure and applied.

Every mathematician has a favorite story illustrating the speed of von Neumann's mind. Once, for example, von Neumann was at a party and his hostess had the nerve to pose a puzzle to him: Two trains are on the same track, heading for a collision. The trains are exactly a mile apart and each is traveling at 30 miles per hour, when a very speedy fly perched on the front of one takes off toward the other train at 60 miles per hour. When the fly lands on the other train it instantly turns around and flies back. It keeps this up, flying from train to train, until it is crushed in the inevitable noisy collision. How far does the fly travel?

Most people, especially those who know a little math, solve this problem by working out the distance the fly travels on each of the legs of its journey and adding them all up. That involves summing an infinite series, which, while not difficult, can be messy and time-consuming—the kind of calculation that most people only tackle with pencil and paper. But there's a trick. First figure out how long before the two trains collide. Since both are traveling at 30 miles per hour and will collide after traversing a half-mile, the crash will occur in one minute. In one minute the fly, which is moving at 60 miles per hour, will travel one mile. Easy.

Almost as soon as his hostess stopped explaining the problem, von Neumann said, "One mile."

"I'm surprised you got it so quickly," she said. "Most mathematicians don't see the trick and use an infinite series to solve this problem, which takes them a few minutes."

"What trick? I solved the problem with an infinite series," von Neumann replied.

Even more maddening, despite his Martian intelligence, von Neumann was a fun guy—charming, a snappy dresser, and the life of the party. "He gave these immense parties, the best ones in Princeton. He loved women and fast cars. He loved jokes, limericks, and off-color stories. He loved noise, Mexican food, fine wines, and money. You just couldn't hate a man like that," Ed Regis wrote in his entertaining book about the Institute for Advanced Study, *Who Got Einstein's Office?*

"In speed and understanding von Neumann was certainly phenomenal," Erdős admitted. "He could understand a proof even far from his own subject very fast." Once Erdős told von Neumann about a proof he had worked out of a theorem in the field of interpolation theory. "It was really not his subject," Erdős said. "He wasn't that interested in it." Von Neumann listened politely and said: "Something seems to be wrong with that proof." After he left, Erdős took another look at the proof and discovered to his surprise that von Neumann had been right. Despite Erdős's respect for von Neumann, the two mathematicians were too different, both in mathematical interests and in personality, ever to collaborate.

After an extremely productive year, Erdős was shocked to learn that his fellowship had not been renewed. Erdős took this rejection as a personal affront, but renewal at the Institute depended on factors beyond the control of even the high and mighty. For example, Leopold Infeld, Einstein's assistant and a well-known physicist in his own right, was also not renewed. Despite his godly status at the Institute, Einstein was unable to obtain even a modest $600 for the 1937–38 academic year for Infeld. "I tried my best," Regis quotes Einstein as having told Infeld at the time. "I told them how good you are, and that we are doing important scientific work together. But they argued that they don't have enough money. . . . I don't know how far their arguments are true. I used very strong words which I have never used before. I told them that in my opinion they were doing an unjust thing. . . . No one helped me."

Melvyn Nathanson, a close friend and collaborator, would recall that Erdős claimed to have been the only person to have been "fired from the Institute of Advanced Study." In those days, Erdős felt, almost everyone got at least one additional year's appointment. The reason he was not renewed is not clear. In part, it might have had something to do with what Nathanson called "a sort of quirky reputation." It is easy to imagine Erdős, arms aflap, full of nervous energy one moment, nodding in lectures the next, scavenging blackboards for problems or hunched over a Go board, not quite fitting into Flexner's Olympian paradise.

The kind of mathematics Erdős loved and pursued so brilliantly also did not fit into the Institute culture. Erdős was not interested in many of the latest developments in mathematics; the math he had mastered as a youth still held inexhaustible treasures in his hands, so why not continue to mine them? Nathanson speculates: "It's as if Paul, when he was young, became an absolute master of certain parts of mathematics, also very beautiful parts of mathematics, and his technique and his imagination in those areas were so strong that he could produce an unending stream of mathematics, never having to go outside very far. And maybe other people who didn't have his depth of imagination or his technical ability needed to learn more mathematics in order to keep generating ideas and new theorems."

And maybe it was just the money. On October 30, 1939, Oswald Veblen, head of the Institute's mathematics department, wrote a generous letter to Herbert Maas, an Institute trustee, beseeching the National Coordinating Committee to provide funds to support Erdős until he could find a more permanent position. In his letter, Veblen described Erdős as "a man of the very highest quality." "We expected him to return to England at the end of his appointment here, but this has been made impossible by the war, although Professor Mordell and other English mathematicians are anxious to do anything that is possible on his behalf," Veblen wrote. That expectation, coupled with the increasing flood of brilliant refugees, might explain why Erdős had not been renewed. Veblen said that Erdős would be a welcome guest at the Institute, and at Johns Hopkins as well, where he had received an honorary fellowship. "Unfortunately," he explained in his letter, "neither Johns Hopkins nor the Institute has any unallocated funds for stipends for the present year." Veblen asked for $1000— more than Einstein had requested for Infeld—which he felt would easily last until some other institution inevitably offered Erdős a job.

The National Coordinating Committee managed to provide $750, which was enough to support Erdős at the Institute for the second term of the 1939–40 academic year. Erdős would treasure the hope of becoming a permanent member of the Institute for more than a decade, before a bitter controversy over what should have been his greatest triumph forced him to admit that it would never be. As the gates of paradise shut behind him, Erdős reluctantly resumed his mathematical journey.

CHAPTER SEVEN

THE JOY OF SETS

FOR the first time in his life, Erdős was penniless. In the months between the expiration of his fellowship at the Institute for Advanced Study and the half-year reprieve Veblen had obtained for him at the beginning of 1940, Erdős depended on the "unlimited credit with his friends" that Vazsonyi had once assured *Anyuka* would be available in his times of need. He managed to survive on loans of fifty dollars or less but could no longer afford to send money to help friends and relatives at home. Erdős would always feel that his favorite aunt in Slovakia could have been saved from the Nazis if he could have sent money for bribes and travel.

Less than a year after he had arrived in the United States, Erdős's skepticism of Chamberlain's appeasement proved to be justified. On September 1, 1939, Hitler attacked Poland, and World War II began. László Alpár, Erdős's politically outspoken friend who had been arrested ten times in Budapest for political activities, fled to France in the hope of continuing his studies in peace. Then the French rounded up left-wing foreigners, including Alpár, and sent them to the Vernet internment camp. Alpár was forced to write all of his letters to his parents in French. Erdős's father, Alpár recalled, "translated the letters I sent to my parents from French to Hungarian, and he also helped them with the translation of their letters to me." Alpár learned of Erdős's whereabouts from Erdős's father, and the friends began to correspond.

Erdős sent Alpár packages and whatever money he could spare. Even more critically, he wrote letters on Alpár's behalf to the French government and eventually managed to secure Alpár's transfer from Vernet to the immigration camp of Les Miles. Unfortunately, Alpár was unable to obtain an immigration visa to the United States, so he remained in Les Miles until the Germans occupied France. He was then moved to the Mirames labor camp, from which he escaped in 1944 to join the French Resistance.

At the beginning of the war mail service between the United States and Hungary, which was allied with Germany, stopped. However, mail still traveled between the United States and France, and between France and Hungary, so Erdős was able to keep in touch with his parents and friends through Alpár. In this way Erdős learned in 1942 that his father had died of a heart attack at the age of sixty-three. The sad news was the last word of his family he would receive during the war.

When his final half-year at the Institute was over, in the summer of 1940, Erdős, with the help of his colleagues at the Institute, was appointed as a Harrison Fellow at the University of Pennsylvania for the 1940–41 academic year. The Institute director, Frank Aydelotte, wrote to a friend of Erdős about the good news and concluded: "So it looks as if he were taken care of for the present and likely to have the opportunity of making a real place for himself in this country."

Whether because of the war or because worrying about the fate of his friends and family distracted him, Erdős published only four papers in 1941. Many mathematicians would consider that a productive year, but for Erdős such output can only be an indication of his extremely depressed state. In Erdősese, to be dead means to not be producing original mathematics. By Erdős's own exacting standards, in 1941 he was near death. The only year of his long career in which Erdős wrote fewer papers was 1934, the year of his first publications.

The Polish mathematician Stanislaw Ulam recalled meeting Erdős for the first time in 1941 when Erdős came to give a colloquium at the University of Wisconsin, in Madison. Despite his already formidable reputation, Ulam was a lowly instructor at the university, waiting out the war in self-described exile, grading exam papers for the Army Correspondence School to supplement his salary.

Ulam, a gregarious man of thirty with a healthy and well-deserved ego, was instantly impressed with the younger Hungarian mathematician. "Erdős was one of the few mathematicians younger than I at that stage of my life," Ulam would write. "In 1941 he was twenty-seven years old, homesick, unhappy, and constantly worried about the fate of his mother, who had remained in Hungary."

Erdős and Ulam quickly forged an intense though intermittent friendship. In his fascinating autobiography, written in 1976, Ulam painted a vivid picture of his friend: "Erdős is somewhat below medium height, an extremely nervous and agitated person. At that time he was even more in perpetual motion than now—almost constantly jumping up and down or flapping his arms. His eyes indicated he was always thinking about mathematics, a process interrupted only by his rather pessimistic statements on world affairs, politics, or human affairs in general, which he viewed darkly. If some amusing thought occurred to him, he would jump up, flap his hands, and sit down again. In the intensity of his devotion to mathematics and constant thinking about problems, he was like some of my Polish friends—if possible even more so."

During Erdős's visit to Madison, the two men did an enormous amount of mathematical work together, "interrupted only by reading newspapers and listening to radio accounts of the war or political analyses." The subject of their collaboration was set theory, perhaps the most abstract branch of mathematics. Erdős would make many important contributions to set theory, but in 1941 he had not yet published any research in the field. Oddly enough, though they would announce some of their findings at an American Mathematical Society meeting, Erdős and Ulam would not publish a joint paper on their work until 1968.

In the 1960s grade school students in the United States were introduced to the rudiments of set theory as part of an idealistic program known as the "New Math." The idea behind New Math was that students would learn math best if they were introduced to the fundamentals lurking beneath the surface, and nothing is more fundamental than set theory. Unfortunately, the set theory taught in grade schools was bowdlerized, stripped of all concepts thought to be too disturbing or difficult for young minds. The New Math version of set theory did not contain the concept of infinity, and set theory without infinity is like Shakespeare without poetry, Cezanne without color. As a boy, Erdős had learned set theory from his father, the X-rated version, filthy with infinities. It was an experience from which he never entirely recovered. In fact, mathematics itself is still reeling from the impact of set theory.

Toward the end of the nineteenth century the German mathematician Georg Cantor decided it was time to tackle the taboo subject of infinities. In 1831 Karl Friedrich Gauss had summed up the attitude of most mathematicians when he expressed his "horror of actual infinities." Gauss's problem was not with infinity *per se*, but with the use of infinity as an object rather than an unobtainable limit. "I protest against the use of infinite magnitude as something completed, which is never permissible in mathematics," he wrote. "Infinity is merely a way of speaking." To Gauss infinity existed only as a limit forever out of reach.

In expressing his dismissal of infinity, Gauss was echoing an opinion that had been generally held practically forever—or at least since Aristotle had rejected the existence of quantities both infinitely large and infinitesimally small more than two thousand years earlier. Cantor would demolish Aristotle's logic and state in 1886: "All so-called proofs against the possibility of actually infinite numbers are faulty, as can be demonstrated in every particular case, and as can be concluded on general grounds as well. . . . the infinite numbers, if they are to be considered in any form at all, must (in their contrast to the finite numbers) constitute an entirely new kind of number, whose nature is entirely dependent upon the nature of things and is an object of research, but not arbitrariness or prejudices."

Cantor showed that not only does the concept of infinity make mathematical sense, but infinities exist in a never-ending hierarchy of increasing sizes. This may sound like a Blakean fantasy, but the genius of Cantor was to prove that towering infinities follow from the sober, irrefutable mathematics of set theory.

A set, as teachers of the New Math would explain, is just a collection of objects—any objects. The Democratic members of Congress, the fish in the sea, the parts in a Boeing 747, all form sets. We say that two sets are the same size if the members in one can be put into a one-to-one correspondence with those of the other. To prove that the set of Musketeers, say, is the same size as the set of Stooges, all you have to do is show that for every Musketeer there is a Stooge and vice versa. One way to do so is:

$$
\begin{array}{ccc}
\text{Moe} & \leftrightarrow & \text{Aramis} \\
\text{Larry} & \leftrightarrow & \text{Porthos} \\
\text{Curly} & \leftrightarrow & \text{Athos}
\end{array}
$$

The number of elements in a set is called its cardinal number. The cardinal number of the set of Musketeers is 3, as is the cardinal number of the set of Stooges.

The same approach works equally well with infinite sets, which can lead to some odd results. The set of all positive integers, {1, 2, 3, . . . } seems to be twice as large as the set of all even numbers, {2, 4, 6, . . . }. Actually, they are the same size, which can be easily seen by setting up a one-to-one correspondence:

$$
\begin{array}{ccc}
1 & \leftrightarrow & 2 \\
2 & \leftrightarrow & 4 \\
3 & \leftrightarrow & 6 \\
4 & \leftrightarrow & 8
\end{array}
$$

And so on. The set of all odd numbers is also the same size, as is the set of all squares or primes. With a little more work it is possible to show that the set of all rational numbers—fractions—is also the same size as the set of all positive integers. That is, it is possible to assign a unique integer to every fraction: The rational numbers can be counted. Cantor called the cardinal number of a countable set— a set that can be put into one-to-one correspondence with the positive integers—$\aleph_0$, read aleph-null. Aleph is the first letter of the Hebrew alphabet, previously unexploited by mathematicians, who had already more or less exhausted the Latin and Greek alphabets in their quest for new typological symbols.

At first glance it might appear that given an endless supply of integers any set could be counted. But Cantor showed that such is not the case. The set of all the decimal numbers—what mathematicians call the real numbers—between 0 and 1, which represents the unbroken continuum of the number line, is uncountable and is infinitely larger than the infinity of the integers. That is, as we shall shortly prove, any attempt at assigning all the real numbers to integers is destined to fail; no matter how hard you try there will always be an infinite number of real numbers left over. Cantor's simple proof of this assertion, using his famous "diagonal argument," is one of the most beautiful and surprising in all of mathematics. After Erdős's father showed him the proof, Paul fell in love with the infinite, and Cantor became his hero. According to Szekeres, Erdős would always regard Cantor's proof to be "the most striking example of a proof coming straight from the celestial book." In those days he closed his letters with "Let the spirit of Cantor be with you," or "C. with you," if he was in a hurry.

Cantor's proof uses the favorite gambit of *reductio ad absurdum*, which by now should be familiar. He assumes that some ingenious soul had managed against all expectations to enumerate all the real numbers between 0 and 1 and produced a table that looks something like this:

$$
\begin{array}{ccl}
1 & \leftrightarrow & .13493358\ldots \\
2 & \leftrightarrow & .85195719\ldots \\
3 & \leftrightarrow & .14159265\ldots \\
4 & \leftrightarrow & .17283845\ldots \\
5 & \leftrightarrow & .04146492\ldots \\
6 & \leftrightarrow & .71582381\ldots
\end{array}
$$

For every integer there is a unique real number between 0 and 1, and for every real number—that is, every possible infinite string of digits following a decimal point—there is an integer.

From this table Cantor constructed a number by following this procedure: for the first digit, take the first digit of the first number in the list; for the second digit, take the second digit of the second number in the list; for the third digit, take the third digit of the third number in the list; and so on. In other words, construct a new number by using the numbers that lie on the diagonal of the table.

1	↔	.13493358...
2	↔	.85195719...
3	↔	.14159265...
4	↔	.17283845...
5	↔	.04146492...
6	↔	.71582381...

In this case, the number Cantor would construct begins, .151863. Next, change each digit of this number into a different digit. It doesn't matter what you change it to as long as it's different. You could, for example, add 1 to each digit, with the exception that 9 becomes 0. Following this rule the new number we construct is .262974.*

The new number is obviously just another decimal between 0 and 1, and since our table is supposed to be complete, it must appear somewhere, corresponding to some integer. In other words, it must already have been counted. But where? To what number does it correspond? It can't be the first number, because the rule by which it was created assures that its first digit is different from the first digit of the first number. The new number can't be the second either, since its second digit is different from the second digit of the second number. In general, the number we constructed can't be equal to the nth number since its nth digit was selected to be different from the nth digit of the nth number. *Therefore, the real number we created cannot be equal to any real number in the table.* Remember: The table was assumed to enumerate exhaustively all the real numbers between 0

---

* To be completely accurate some care must be taken to avoid a small problem that arises because, for example, .2499 ... (with an infinite number of 9's) and .2500 ... (with an infinite number of 0's) are actually the same number. This problem is avoided by never changing a digit to a 9 or a 0.

and 1, but we found a real number that is, to paraphrase Samuel Goldwyn, included out. Using the diagonal digits of all the real numbers listed in the table, we built a real number that does not appear in the table, although the table was supposed to be complete. This proves that, try as one might, any attempt to construct a one-to-one correspondence between the real numbers and the integers is doomed to failure. The infinity of the real numbers—the so-called continuum—is larger than the infinity of the integers.

Cantor went on to show how to construct an entire hierarchy of ever larger infinities, a world of incredible richness for mathematicians to explore. His development of set theory also paved the way for paradoxes that have revealed fissures in the foundation of mathematics.

Groucho Marx once said that he wouldn't want to belong to any club that would have him as a member. Sets aren't as choosy; a set can be a member of itself. For example, "the set of all objects that can be described in a hundred words or less" is itself an object that can be described in fewer than a hundred words, and thus must have itself as a member. So far so good. Some sets are members of themselves and some are not (for example, the set of all Hungarian mathematicians is not a Hungarian mathematician); all sets must be one or the other. Shortly after the turn of the century Bertrand Russell demonstrated how this apparently unexceptionable statement leads to a paradox that threatened to destroy mathematics.

Consider a peculiar set that we'll call R in honor of Russell. By definition, the set R contains all sets that do not have themselves as members. Is R a member of itself? Clearly not, since by definition R is the set of all sets that *do not* have themselves as members. But then, if R does not have itself as a member, then (again by the definition of R) it *must* be a member of R.* As Russell observed, "each alternative leads to its opposite and there is a contradiction." The logic is impeccable, the result catastrophic.

"I felt about the contradictions much as an earnest Catholic must

---

* The essence of Russell's Paradox is captured in the famous riddle of the town barber: In a small town where all the men are clean-shaven, the barber shaves everyone who does not shave himself. Who shaves the barber?

feel about wicked Popes," Russell said. He wrote a letter describing the paradox to a logician named Gottlob Frege, who was in the process of completing a huge work on the foundations of arithmetic. Frege made liberal use of the set concept, including sets that contained themselves as members, in his book. Embarrassingly, the book also contains, according to the historian of mathematics E. T. Bell, "a considerable use of more or less sarcastic invective against previous writers on the foundations of arithmetic for their manifest blunders and manifold stupidities." Frege's honesty, however, cannot be faulted. The second volume of his magnum opus closes with his shaken acknowledgment of Russell's bombshell: "A scientist can hardly encounter anything more undesirable than to have the foundation collapse just as the work is finished. I was put in this position by a letter from Mr. Bertrand Russell when the work was in press."

The foundations of Russell's life were also severely shaken by his discovery. As has been mentioned, ever since Russell had first gazed into Euclid's *Elements* at the age of eleven, he claimed that "mathematics was my chief interest, and my chief source of happiness." When he was sixteen and miserable, only the desire to learn more mathematics prevented him from committing suicide. With that in mind it is not hard to understand the disillusion and despair Russell experienced upon discovering the shattering paradox that would always be associated with his name. Afterward he "turned aside from mathematical logic with a kind of nausea," leaving to others the task of cleaning up the mess he created.

The great German mathematician David Hilbert, who considered Cantor's set theory to be "the finest product of mathematical genius and one of the supreme achievements of purely intellectual human activity," was no less disturbed, though he was not as easily driven to despair as Russell. "[T]he present state of affairs ... is intolerable," he admitted in 1925, during a famous lecture on the subject of infinity. "Just think, the definitions and deductive methods which everyone learns, teaches, and uses in mathematics, the paragon of truth and certitude, lead to absurdities! If mathematical thinking is defective, where are we to find certitude?" Hilbert exhorted his audience to take heart and appealed to their pride. "Let us remember that *we are mathematicians* and that as mathematicians we have often been in precarious situations," he thundered. "No one will drive us out of this paradise that Cantor has created for us."

Hilbert was right, though the terms of the victory might not have entirely satisfied him. Mathematicians have resolved the paradoxes of set theory, and Cantor's paradise is not lost. The price, in part, has been to allow uncertainty to creep into mathematics via Gödel's theorem. Using an ingenious twist on Cantor's diagonal method, Gödel proved that some mathematical questions are simply undecidable within the system in which these questions are formulated. That news can be either disappointing (some problems can't be solved) or exhilarating (mathematical systems can be continually expanded, meaning there is no end to mathematics), depending upon your point of view. Erdős never worked directly in the field of mathematical logic, though he admired Gödel's proof. According to Szekeres, Erdős "was greatly thrilled when undecidable statements turned up later in quite down-to-earth graph problems of his."

Erdős would make important contributions to Cantor's paradise of set theory through his extension of Ramsey theory—the field he had pioneered with George Szekeres during his early Budapest years—into the realm of the transfinite (the term mathematicians use to refer to Cantor's infinity of infinities) and through his work on what are called inaccessible cardinals. Most of this work was done throughout his life with Richard Rado and András Hajnal, a mathematician Erdős would meet in Budapest in the late 1950s.

For Erdős, publication was a form of courtesy, the ultimate act of sharing, but sometimes, as in his collaboration with Ulam, for whatever reason, interesting results of his did not find their way into print. Erdős was acutely aware of the negative example of Gauss, who published only works that he considered finished masterpieces. "I get no pleasure from incomplete solutions," Gauss said, "and work in which I have no joy is torture to me." Gauss even adopted a seal depicting a fruit tree, bare but for a few large, beautiful fruit, and the motto *"Pauca sed matura"* ("Few but ripe"), anticipating, perhaps, Paul Masson's boast of selling no wine before its time. The jottings in Gauss's notebooks, discovered years after his death, would have advanced mathematics by decades had they been published. Even more infuriating and destructive was Gauss's habit of telling mathematicians who came to him with new results that he had anticipated them by decades; praising them would be praising himself, which Gauss, being a modest man, could not do. János Bolyai, who did pioneering work in non-Euclidean geometry, was devastated when Gauss informed him that he had already covered the same ground. So, when a young Indian mathematician told Erdős about his latest discovery, Erdős did not have the heart to reveal that he and Ulam had discovered the same result years earlier but had not published it. Erdős praised the work instead, and urged the young man to publish. Several years later the Indian mathematician (Erdős claims to have forgotten his name, which, given his elephantine memory, seems less likely than tactful) found out about Erdős's earlier results and asked him, "Why didn't you tell me?"

"I didn't want to imitate Gauss in this regard," Erdős replied.

Ulam, who had a talent rare among pure mathematicians for solving real-world problems, was recruited in 1942 by his friend John von Neumann to join the secret Manhattan Project, which was developing the atomic bomb at Los Alamos, New Mexico. Ulam was assigned to the critical but messy problem of calculating the hydrodynamics of the implosion needed to crush together the fissionable material in a bomb to form the critical mass. Von Neumann and the other brilliant mathematicians and physicists tried to find an ingenious theoretical simplification that would allow them to calculate the details with pencil and paper. Ulam, despite his abstract background, understood that the best approach was to use "simpleminded brute force— that is, more realistic, massive numerical work." Ulam contacted his friend Erdős to come to Los Alamos to pitch in. Erdős was willing, and following Ulam's advice he wrote to "Professor Edwards" (his interpretation of Edward Teller). Unfortunately, Erdős told Teller that he might want to return to Hungary after the war, which was enough for Teller to deny the project the benefit of Erdős's talent.

Erdős was invariably direct in matters personal as well as political, as Ulam would discover. In 1945, when the war was over, Ulam accepted a position at the University of Southern California in Los

Angeles. Shortly after beginning at USC, Ulam was stricken by a sudden and intense headache, accompanied by a frightening numbness creeping from his chest toward his chin. "I remembered suddenly Plato's description of Socrates after he was given the hemlock," Ulam recalled. His wife summoned a doctor, and Ulam was rushed to the hospital.

For some reason that the doctors were unable to explain, Ulam's brain had begun to swell dangerously. An emergency operation relieved the pressure, and in a few days Ulam was apparently fine. To check his mental recovery the doctor asked him, "What is 13 plus 8?" "The fact that he asked such a question embarrassed me so much that I just shook my head," Ulam wrote. Concerned, the doctor then asked him what was the square root of 20. "About 4.4," Ulam replied. The doctor was silent. "Isn't it?" Ulam asked, a little concerned. The doctor laughed and said, "I don't know." Simple arithmetic convinced the doctor that Ulam was all right, but Ulam, who relied more heavily on the flawless functioning of his brain than most people, needed more convincing.

On the day he was to be released from the hospital Erdős appeared in Ulam's room, carrying a suitcase. "Stan," he said, "I am so glad to see you are alive. I thought you were going to die and that I would have to write your obituary and our joint papers." Ulam was gratified by Erdős's obvious pleasure in his survival but a little concerned that his friends had given him up for dead. Having no immediate plans, and seeing that Ulam was about to check out, Erdős said, "You are going home? Good, I can go with you."

During the entire trip to Ulam's home on Balboa Island, Erdős talked mathematics. As Ulam recalled, "I made some remarks, he asked me about some problems, I made a comment and he said: 'Stan, you're just like before.'" Ulam was pleased but still unconvinced. When they arrived at Ulam's home Erdős instantly challenged Ulam to a game of chess. Only after beating Erdős two games in a row did Ulam start to accept that his brain, which for a few hours on an operating table had been more open than even Erdős's, was undamaged. After two weeks of Erdősian challenges, Ulam was convinced that his mental abilities were undiminished.

The mathematical discussions they had during walks on the beach completed the proof of Ulam's recovery. On one walk Erdős stopped, as he always did, to say hello to a child. "Look, Stan! What a nice epsilon." The child's mother, who was standing nearby, was a beautiful young woman, and Ulam could not resist replying, "But look at the capital epsilon," causing Erdős to blush with embarrassment.

That Erdős would materialize for a visit did not strike Ulam as the least bit surprising. During those years Erdős survived on fellowships that lasted no more than an academic year, at places like Purdue University and the University of Michigan, and on tiny lecture fees and the kindness of colleagues. It never really mattered to Erdős where he was since, like Einstein, he carried his laboratory with him and worked constantly, oblivious to his surroundings. He once was stopped after midnight not far from the Purdue campus by some police officers who thought he looked suspicious. "What are you doing?" they asked him. "I am thinking," he explained. "You are," the officer admitted reasonably and walked on.

The authorities were not always so reasonable or so easily mollified. In August 1941 Erdős was in New York visiting some mathematicians when he briefly became, as the *Philadelphia Inquirer* put it, "the center of a spy scare." With some time to kill, Erdős had decided to accompany the British mathematician Arthur Stone and a Japanese mathematician named Shizuo Kakutani on a brief tour of Long Island. They drove out to the eastern tip of the Island and checked into a small motel. That night Erdős noticed an interesting light beckoning to him in the distance. The next day at his urging the three went in search of the light. The road they followed, which branched off the highway and narrowed as it climbed a tract of land called Barren Hill, should have been a warning. So should have been the "No Trespassing" sign.

Kakutani snapped a few photographs of Stone and Erdős against a background of mysterious buildings that was a secret radar installation. Two radio operators would later report having seen one of the trespassers sketching the 200-foot tower. They were spotted by a guard, who asked them to leave, so they got back into the car and drove to a diner for some lunch. As they were paying their check, the mathematicians were approached by a pair of "enormous" policemen who arrested them and hauled them to the station house for questioning. The guard had reported that "three Japanese had taken pictures of the installation and then departed in a suspicious hurry." Three foreign nationals of any kind would have been enough to alarm even the most phlegmatic guard. In those years good citizens were taught to be ever vigilant; even taking photographs of New York harbor from the rail of the Staten Island ferry was prohibited, and fishing boats trolling the waters off Long Island's south shore were required to have special licenses, and their passengers had to carry special identification cards issued by the Coast Guard.

The FBI was summoned, and the three suspected spies were questioned separately. Oswald Veblen at the Institute for Advanced Study vouched for Erdős, and other colleagues helped clear Stone and Kakutani. The police finally released them later that night, with apologies, but could not forbear pointing out that the whole incident could have been avoided. How could they have missed the "No Trespassing" sign?

"You see, I was thinking about mathematical theorems," Erdős explained.

As a Japanese citizen Kakutani found it difficult to work in the United States during the war. One of Kakutani's papers was rejected by the *Duke Mathematical Journal* because "the author is Japanese and did not show his disagreement with the Japanese government." It probably had not occurred to Kakutani that the Japanese government held any opinions at all regarding the decomposition of reals into Hamel bases, or if they did, that these might be abhorrent to the United States. Consumed like Erdős with homesickness and a concern for his mother, who was ill, Kakutani managed to return to Japan in 1942 aboard a Swedish ship. His later return to the United States was probably facilitated by his having been previously cleared by the FBI after the Long Island incident.

With no word of the fate of his friends and family in Hungary, Erdős could only imagine the worst. In March 1944 Germany occupied Hungary and began its systematic extermination of Hungarian Jewry. Adolf Eichmann arrived at the head of the Nazi forces and immediately established a Jewish ghetto. Almost all Hungarian Jews living outside of Budapest perished in Auschwitz during the last year of the war. Of those in the Budapest ghettos at least half survived, many through the intervention of the Swedish diplomat Raoul Wallenberg.

Erdős's worst fears were calmed when he received a telegram from a friend in Romania in 1945 with the news that his mother was alive. A few months later he received his first letter directly from his mother.

Erdős's joy in his *Anyuka*'s survival, however, was diminished by the realization of the tragic death toll the Nazis wreaked among the rest of his family and friends. Four out of five of Erdős's aunts and uncles, and the husband of the surviving aunt, perished in the Holocaust. Two of Erdős's close friends from the Anonymous group, Géza Grünwald and Dezso Lázár, were killed by the Nazis, as were the editor of *KöMal*, the journal that had brought them together, Andor Faragó, and his sons. Dénes König, in whose class Erdős had learned graph theory, committed suicide when ordered to move to the ghetto. Not until 1948, more than forty papers later, would Erdős return to Budapest and his beloved *Anyuka*.

CHAPTER EIGHT

THE PRIMES OF
DR. PAUL ERDŐS

I N the years following the war Erdős's papers spanned an increasingly broad range of disciplines. When he met Mark Kac, Erdős knew very little about probability theory, but within a few years he had become a leading expert in the field; he wrote papers on combinatorics, graph theory, geometry, and interpolation, which concerns estimating functions from a few given values. But the majority of Erdős's papers were still largely devoted to his first love, the Queen of Mathematics: the theory of numbers.

Ever since he was ten years old and his father had proved to him that the gaps between successive primes can be arbitrarily large, Paul had been fascinated by the patterned disorderliness of the prime numbers. The primes seem to be almost randomly distributed like oases across the vast desert of composite numbers. They obey no known formula. Finding new primes is a laborious process of exhaustive exploration best performed by planetary networks of computers. And yet if individual primes are ignored, primes in the aggregate obey a simple and beautiful law known as the Prime Number Theorem.

The proof of the Prime Number Theorem was found in the final years of the nineteenth century and remains one of the supreme achievements of mathematics. Erdős would have learned of the theorem while still a young student, but he would never find the classic proof entirely satisfactory. The proof was extremely difficult to follow, but that was not in itself a problem. More troubling was that the proof was not "elementary," a word that to mathematicians has nothing to do with difficulty. An elementary proof uses only the procedures of classical number theory together with the properties of integers and real numbers. The known proofs of the Prime Number Theorem employed such concepts as continuous functions and complex numbers, which are hard to relate intuitively to the properties of integers. Even as a young man Erdős found it hard to accept that the proof of the Prime Number Theorem that was inscribed in The Book could not possibly depend on them. While growing up, Paul told his friends that he dreamed of someday coming up with a truly elementary proof of the Prime Number Theorem, something many people considered impossible. In 1948 Erdős would realize that boyhood ambition when he found the elementary Book proof. It would be one of the crowning triumphs of his career and would bring him honors and fame. The proof would also embroil Erdős in the only real controversy of his life.

T H E first person to construct a table of prime numbers was Eratosthenes, who lived in Greece during the third century B.C. Instead of laboriously testing numbers one at a time, he invented a clever short cut for quickly sifting large blocks of numbers, known as the Sieve of Eratosthenes. Say you want to construct a table of the primes amid the first fifty numbers. First write down the first fifty integers:

1	2	3	4	5	6	7	8	9	10
11	12	13	14	15	16	17	18	19	20
21	22	23	24	25	26	27	28	29	30
31	32	33	34	35	36	37	38	39	40
41	42	43	44	45	46	47	48	49	50

Then cross out 1, which—by definition—is not a prime. Two is a prime, so leave it alone, but cross out every second number after 2. Those are multiples of 2, and hence are not prime.

~~1~~	2	3	~~4~~	5	~~6~~	7	~~8~~	9	~~10~~
11	~~12~~	13	~~14~~	15	~~16~~	17	~~18~~	19	~~20~~
21	~~22~~	23	~~24~~	25	~~26~~	27	~~28~~	29	~~30~~
31	~~32~~	33	~~34~~	35	~~36~~	37	~~38~~	39	~~40~~
41	~~42~~	43	~~44~~	45	~~46~~	47	~~48~~	49	~~50~~

The first table entry after 2 that is not eliminated is 3, which must therefore be a prime. Cross out all multiples of 3 by counting off every third number.

~~1~~	2	3	~~4~~	5	~~6~~	7	~~8~~	~~9~~	~~10~~
11	~~12~~	13	~~14~~	~~15~~	~~16~~	17	~~18~~	19	~~20~~
~~21~~	~~22~~	23	~~24~~	25	~~26~~	~~27~~	~~28~~	29	~~30~~
31	~~32~~	~~33~~	~~34~~	35	~~36~~	37	~~38~~	~~39~~	~~40~~
41	~~42~~	43	~~44~~	~~45~~	~~46~~	47	~~48~~	49	~~50~~

And so on. The next entry not crossed out is 5, so it is prime. Keep it and eliminate every fifth entry. You have to continue this only until you reach an entry that is bigger than the square root of the number of entries in the table. In this case, since the table has fifty entries, you are through after crossing off 7, since the square root of 50 is greater than 7 and less than 8. The reason this works is that any number that is not prime must be the product of two or more primes. No two of these prime factors can be greater than the square root of the number, because their product would be larger than the number. For example, no two distinct prime factors of 100 can be bigger than 10, the square root of 100, since the product of any two numbers greater than 10 is greater than 100. So at least one of the prime factors of every entry in the table must be smaller than the square root of the number of table entries. By the time you have crossed out all multiples of numbers greater than the square root of the number of table entries you've eliminated all nonprime numbers in the table. Here is the final result of Eratosthenes' sieving procedure on the first fifty numbers:

~~1~~	2	3	~~4~~	5	6	7	8	9	~~10~~
11	~~12~~	13	~~14~~	~~15~~	~~16~~	17	~~18~~	19	~~20~~
~~21~~	22	23	24	25	26	~~27~~	28	29	~~30~~
31	32	33	34	35	36	37	38	39	~~40~~
41	42	43	44	45	46	47	48	49	~~50~~

After Eratosthenes, the compiling of tables of prime numbers became an activity of almost obsessive religious devotion to the Queen of Mathematics. In 1776 Antonio Felkel tried to purchase immortality by compiling a table of the prime factors of all the numbers up to two million. The first volume of his table, containing the factors of all numbers less than 408,000, was published but was not the literary success its author had hoped for. All but a few volumes were subsequently used to make cartridges in the Turkish war. The Viennese imperial treasury, which had paid for the unsuccessful printing of the first volume of Felkel's manuscript, kept the remaining unpublished pages, it would seem, out of spite. Undeterred, Felkel recalculated the expropriated pages and extended his previous effort up to 2,856,000.

The Polish mathematician Yakov Kulik is certainly the most heroic and tragic of the prime fetishists. On the theory that everyone should have a hobby, Kulik devoted all his leisure hours for twenty years to compiling a table of the prime factors of all numbers less than one hundred million. After Kulik's death the residue of his life's work, all eight volumes and 4,212 pages of it, was entrusted to the library of the Royal Academy of Vienna, where it can be found today minus the second volume, 12,642,000 to 22,825,800, which has been lost. The loss, while tragic on a human scale, was less important mathematically. Examination of the first volume uncovered so many errors that Kulik's effort was all but worthless.

When Karl Friedrich Gauss was a boy of fifteen, he examined a table of prime numbers smaller than 102,000 compiled by Johann Lambert, looking for patterns. Mathematics is often said to be the highest expression of pure reason, but Gauss would stress that it is also a science of the eye. By that he meant mathematics comes from close observation of the properties of numbers, shapes, and structures. Mathematicians spend a lot of their time tinkering, looking for regularities and oddities upon which to base their conjectures and proofs. The Indian mathematician Ramanujan spent countless hours filling pads with arithmetical calculations that would inspire his theories. "Ramanujan was doing what great artists always do—diving into his material," his biographer Robert Kanigel wrote. "He was building an intimacy with numbers, for the same reason that the painter lingers over the mixing of his paints, or the musician practices his scales."

The young Gauss decided to step back from the apparent chaos of the primes, like a painter stepping back from her easel to evaluate a work in progress, to see if he could discern any larger patterns at work. Gauss divided the natural numbers into intervals of 1,000 and, using Lambert's table, counted up the number of primes in each interval. The trick worked; viewed from a distance the distribution of primes becomes orderly. Look at two graphs showing the number of primes less than x, a function mathematicians call $\pi(x)$, read "pi of x." The first graph is a close-up, focusing on the numbers from 1 to 100. The second pulls back to show the numbers up to 1,000.

The function, which is so jagged in the first diagram, looks remarkably smooth when viewed from a distance. Using numbers gleaned from Lambert's table, Gauss was able to guess a remarkably simple and precise law that described the distribution of the primes. Gauss's formula accurately captures the slow but inexorable thinning out of

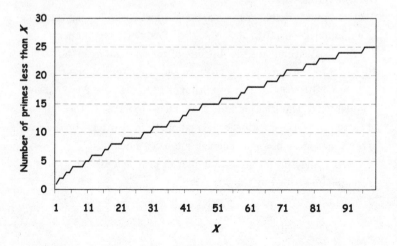

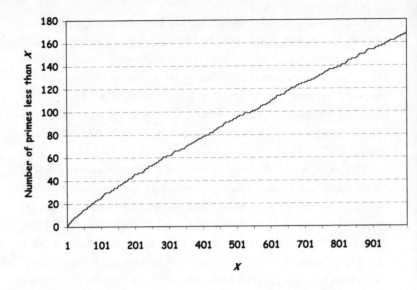

the primes. Primes account for 25 percent of the numbers less than 100 and about 17 percent of the numbers less than 1,000. Of the first million integers, only about 8 percent are prime, and the decline continues, slowly and inevitably. The percentage of numbers less than a trillion that are prime drops to only 4 percent. This last number, by the way, is not the life work of one of Kulik's mad heirs, but was arrived at by a high-speed computer running a very efficient prime-counting program.

Gauss guessed that a mathematical function known as the logarithm could be used to describe the slowly waning percentage of primes. Logarithms are the flip side of exponential growth; 1,000 is 10 raised to the third power (10^3), so the *logarithm* of one thousand to the *base* 10 is three. The logarithm of 16 to the base 2 is 4, since 2^4 is equal to 16. The Richter scale that is used to measure the intensity of earthquakes is logarithmic; an earthquake measuring 5 on the scale is ten times more powerful than an intensity 4 quake.

Gauss conjectured that the number of primes less than x, or $\pi(x)$, is roughly described by the formula

$$\pi(x) = x/\log(x).$$

In this formula log(x) denotes the natural logarithm preferred by mathematicians, which is taken to the base e, or roughly 2.718 (the natural logarithm is roughly .4343 times the base 10 logarithm most people learn in school). Gauss's contemporary Adrian-Marie Legendre independently guessed at the approximate form of the law. A quick comparison shows how well this formula represents the actual behavior of prime numbers.

x	$\pi(x)$ The actual number of primes less than x	$x/\log(x)$ The estimate of $\pi(x)$ given by the Prime Number Theorem
1,000	168	145
10,000	1,229	1,086
100,000	9,592	8,686
1,000,000	78,498	72,382
10,000,000	664,579	620,420
100,000,000	5,761,455	5,740,304

Throughout his life Gauss continued to check his observed law whenever he could, sometimes turning to the latest tables of primes for guidance but more often relying on his awesome calculating abilities. In 1849 Gauss described his "youthful investigations" into the distribution of primes to the astronomer Johann Franz Encke. "I have since (since I lacked the patience to go through the whole series systematically) often used a spare quarter of an hour to investigate a thousand numbers here and there; at last I gave it up altogether, without ever finishing the first million." From this spare-time tinkering Gauss was able to report to Encke that "The thousand numbers between 101,000 and 102,000 bristles with errors in Lambert's supplement; in my copies I have crossed out seven numbers that are not primes, and in return put in two that were missing."

Gauss offered no proof of his conjectured formula for the number of primes less than a given number, and for many years nobody else could prove it either. In 1852, the Russian mathematician Chebychev was the first to make any progress toward a proof of the Prime Number Theorem, which is the name that has since been applied to Gauss's conjecture. Chebychev's theorem was an extension of his proof of Bertrand's postulate, the same theorem that was the subject of the nineteen-year-old Erdős's first important piece of original work. Recall that Chebychev's theorem can be summarized by the heroic (in spirit if not in prosody) couplet:

> Chebychev said it and I'll say it again,
> There's always a prime between n and $2n$!

In other words, between any number and its double, there is always at least one prime. Chebychev's theorem gives a clue to the distribution of primes. In fact, Chebychev was able to go a lot farther than Bertrand's postulate, but his effort fell short of a proof of the Prime Number Theorem. A contemporary of Chebychev, one who had also failed to find the elusive proof of the Prime Number Theorem, exclaimed that "we shall probably have to wait [for a proof] until someone is born into the world so far surpassing Chebychev in insight and penetration as Chebychev has proved himself superior in these qualities to the ordinary run of mankind."

Gauss's most famous student, Bernhard Riemann, took the next important step toward a proof of the Prime Number Theorem. In awarding Riemann his doctorate, Gauss had praised his "gloriously fertile originality." In his brief career—he died at the age of forty— Riemann would prove himself to be Gauss's worthy successor with his development of what are today known as Riemann integrals and his explorations of the geometry of curved spaces that became a vital ingredient of Einstein's theory of gravitation. In a paper written in 1859 Riemann showed that the problem of counting the prime numbers, essentially a problem of arithmetic, could be attacked by examining the properties of a mathematical object now known as the Riemann zeta function. It was a brilliant leap of intuition, but in taking it Riemann introduced concepts from other branches of mathematics that had previously seemed to have no place in number theory. The Riemann zeta function is not an elementary concept. Erdős believed that its use in proofs of the Prime Number Theorem was unnecessary and in fact rendered the underlying reason for the distribution of prime numbers obscure. To understand the reason for Erdős's belief requires a closer examination of Riemann's peculiar creation.

The Riemann zeta function is one of the most famous examples of what mathematicians call a function of a complex variable. Functions are mathematical machines that take a number as input, chew on it, and spit out another number. The numbers that a function like Riemann's function chew on and spit out are so-called complex numbers. Complex numbers consist of two parts, one of which is a familiar real number and the other a so-called imaginary number.

In reality, all numbers are imaginary. We can give our true love five golden rings, but not the number five. Erdős's pride in his childhood discovery of the negative numbers is justified; what, after all, does it mean to say you have negative three geese a-laying? The Greeks did not believe in negative numbers. To them an equation like

$$x + 3 = 0$$

did not have a solution. It took hundreds of years to accept that it made sense to talk about negative numbers. Imaginary numbers similarly arose when people tried to solve equations like

$$x^2 + 1 = 0.$$

In other words, what number when squared equals negative one? A positive number times a positive number is positive, so no positive number would do. A negative number times a negative is also a positive number, so negative numbers do not work either. If not negative or positive, then what? It was grudgingly admitted that a new kind of number was needed. For reasons probably more spiteful than existential, the new kind of number was called imaginary. With the addition of imaginary numbers—and complex numbers, which are the sum of real and imaginary numbers—equations like the one above could be solved, and an incredible amount of beautiful and useful new mathematics followed. Today complex numbers are routinely used by the ultimate realists, engineers, to help them design everything from cell phones to suspension bridges.

Despite their usefulness in other branches of mathematics, it came as a bit of a surprise that complex numbers should invade the domain of the integers. Riemann showed that a thorough knowledge of the properties of the zeta function would lead to results in the theory of prime numbers stronger than any known. In his paper Riemann made six conjectures about the zeta function that revealed a profound intuitive grasp of a complex domain that continues to astound mathematicians. Five of Riemann's conjectures have been proved, but the last, known as the Riemann hypothesis, remains an open question, perhaps the most important unsolved problem in mathematics.

David Hilbert, who listed it as one of the outstanding problems in mathematics, once remarked that if he was wakened after a thousand-year nap, the first question he would ask would be, "Has the Riemann hypothesis been solved?" Hilbert's biographer, Constance Reid, recounts an anecdote that, while perhaps apocryphal, shows the depth of Hilbert's obsession with the Riemann hypothesis. Hilbert once had a student, or so the story goes, who came to him with a proof of the Riemann hypothesis. Hilbert was impressed with the effort but after careful study found a fatal error. Soon afterward the student died, and Hilbert was asked by his friends to give a funeral oration. Hilbert started by saying all the proper things, regretting the loss of one so young and so talented and so on. He mentioned the student's attempted proof, which while flawed might some day lead to a solution of the famous problem. "In fact," said Hilbert enthusiastically, standing in the rain before the dead student's grave, "let us consider a function of a complex variable . . ."

G. H. Hardy made important strides toward the solution of the Riemann hypothesis, though he was never able to solve it. He even used it in an elaborate game he played with God. Hardy was afraid of sea travel, so before he boarded a ship to cross the North Sea after visiting with his mathematical friend Harald Bohr (younger brother of Niels Bohr) in Denmark, as a bit of travel insurance Hardy would mail Bohr a postcard announcing that he had solved the Riemann hypothesis. Hardy was convinced, Reid reports, "that God—with whom he waged a very personal war—would not let Hardy die with such glory."

It took almost forty years for mathematicians to master the prickly zeta function sufficiently to prove the Prime Number Theorem. In 1896 Jacques Hadamard and Charles Jean Gustave Nicolas de la Vallée Poussin finally proved the Prime Number Theorem, confirming the young Gauss's astute guess. Subsequent generations of mathematicians would labor at improving Hadamard's and Poussin's difficult proof and illuminating the properties of Riemann's zeta function, though the Riemann hypothesis remains unsolved. But no elementary proof was found, and many mathematicians thought such a proof to be unlikely. "We have certain views about the logic of the theory," Hardy wrote. "We think that some theorems, as we say 'lie deep' and others nearer to the surface. If anyone produces an elementary proof of the prime number theorem, he will know that these views are wrong, that the subject does not hang together in the way we have supposed, and that it is time for the books to be cast aside and for the theory to be rewritten."

Erdős and a few other mathematicians viewed remarks like Hardy's to be more of a challenge than a warning. Erdős had declared his devotion to elementary methods with his remarkable proof of the Chebychev theorem in 1931, and continued to champion them throughout his life. In Erdős's hands elementary techniques, which to many mathematicians seemed to have been played out, continued to yield unexpected and beautiful results, proofs ready for publication in The Book. Many of his theorems concerned the distribution of primes, but his goal of finding an elementary proof of the Prime Number Theorem remained elusive.

During World War II mathematicians in Europe were unable to communicate with their colleagues in the United States. After the cessation of hostilities those who could travel journeyed to Europe to see what their friends had been up to. Paul Turán, who had managed to survive the war and its aftermath in Budapest, was invited to spend six months at the Institute for Advanced Study. Overjoyed at the prospect of seeing his friend again, Erdős came down from Syracuse University, where he was a Visiting Resident Professor, to meet Turán when he arrived in New York. Over the next six months Erdős made frequent visits to the Institute to work with Turán on papers on the distribution of roots of polynomials, which are still cited today.

After the war Hermann Weyl, a professor at the Institute, had gone on an exploratory expedition to Europe. He learned of a young Norwegian mathematician named Alte Selberg, who had published some beautiful articles in analytic number theory in an obscure Norwegian journal. Weyl must have felt like a baseball scout coming across a pitcher hurling hundred-mile-an-hour fastballs in a dusty rural sandlot. He quickly signed up Selberg and whisked his new protégé back to Princeton, New Jersey, to play in the big leagues.

Selberg did not disappoint Weyl's expectations. In his approach to creating mathematics Selberg was almost the opposite of Erdős. He was a quiet man, something of a loner, who seldom wrote collaborative papers. Like Erdős, he was a wizard at applying elementary methods to crack difficult number theory problems. In May 1948 Selberg wrote a paper in which he gave an elementary proof of the Dirichlet Theorem, which, next to the Prime Number Theorem, was the outstanding challenge to the power of elementary methods. Dirichlet theorems concern the occurrence of primes in arithmetic progressions. An arithmetic progression is a sequence of numbers that are equally spaced, like 3, 5, 7, 9, 11, ... In this arithmetic progression the terms 3, 5, 7, and 11 are prime. Dirichlet proved in 1837 that any arithmetic progression whose terms do not all share a common factor contains an *infinite* number of primes. So the arithmetic progression 17, 22, 27, 32, . . . contains an infinite number of primes, while the arithmetic progression 5, 10, 15, 20, . . . does not, since in this case each term is a multiple of 5.

Selberg had gotten to know and like Turán and told him about his proof of the Dirichlet Theorem. Selberg was planning a trip to Canada in July, and since Turán knew that he'd probably be gone by the time Selberg returned, he asked to see Selberg's notes on the proof. "I not only agreed to do this, but as I felt very much attached to

Turán I spent some days going through the proof with him," Selberg later wrote to Weyl. Selberg also threw in a little bonus, a glimpse at an eloquent equation he had discovered in March that became known as the "fundamental formula." "I did not tell him the proof of the formula," Selberg wrote, "nor about the consequences it might have and my ideas in this connection."

What Selberg had suspected and had not told Turán was that his fundamental formula might be the key to an elementary proof of the Prime Number Theorem. It was easy to show that Selberg's formula was a consequence of the Prime Number Theorem, but that is not how he found it; Selberg had derived his formula using entirely elementary methods. Since the fundamental formula could be proved *from* the Prime Number Theorem, it seemed possible to Selberg that the reverse might be possible: The fundamental formula could *lead to* an elementary proof of the Prime Number Theorem. Selberg told none of this to Turán and left the Institute for nine days.

During Selberg's absence Turán gave a small, informal seminar on the proof of the Dirichlet Theorem at the Institute. His audience consisted of Erdős, Ernst Straus, who then was Einstein's assistant, and Saravadam Chowla. Both Straus and Chowla would later become Erdős collaborators. "After the lecture," Straus would write, "there followed a brief discussion of the unexpected power of Selberg's inequality [the fundamental formula]." Erdős saw right away that Selberg's formula might imply the Prime Number Theorem. The first step, he felt, was to prove an intermediate theorem of his own that intuition told him was a consequence of Selberg's work. Erdős sensed that this intermediate theorem, which stated roughly that the ratio of successive primes approaches one as the primes grow larger, would bring him close to his ultimate goal of proving the Prime Number Theorem. As usual, he got right to work.

When Selberg arrived back at the Institute he was surprised, though not displeased, that Turán had given a seminar on his work on the Dirichlet Theorem. "I had of course no objection to this, since it concerned something that was already finished from my side, though it was not published. In connection with this Turán had also mentioned, at least to Erdős, the fundamental formula. This I don't object to either, since I had not asked him not to tell this further."

Selberg may not have felt he had the right to object, but he was not happy that Erdős had seized on his results with such enthusiasm. For Erdős the purpose of mathematics—the purpose of life—was to prove and conjecture, and to do so as rapidly as possible. A mathematical theorem once discovered became the property of everyone, and Erdős felt duty-bound to follow its consequences wherever they led. His years of legendary collaboration were still ahead of him, but even so, of the 133 papers he had written before 1948, fifty-two were done in collaboration. Many mathematicians enjoyed Erdős's excited, social style of doing mathematics. And then there were those like Selberg, who preferred to work alone and at their own pace; to them Erdős's aggressive approach to doing mathematics could seem pushy and rude. Or at least overwhelming.

Erdős bumped into Selberg outside of Fuld Hall at the Institute on Thursday afternoon, shortly after Selberg had returned from Canada. Erdős told Selberg about the intermediate theorem he was trying to prove. "I became rather concerned that Erdős was working on these things," Selberg wrote to Weyl shortly after the incident. Selberg tried to discourage Erdős by telling him that he doubted whether the fundamental formula implied the intermediate theorem Erdős was trying to prove, and would probably not yield an elementary proof of the Prime Number Theorem. Selberg even told Erdős that he had constructed a counterexample, the mathematical equivalent of a deal-breaker. But, as he would later admit, Selberg's supposed counterexample was deliberately misleading; Selberg had neglected to tell Erdős about certain underlying assumptions that nullified its damaging effects. "This attempt to throw Erdős off the track (clearly not succeeding!) is somewhat understandable given my mood at the time," Selberg would explain in a letter written almost fifty years later.

The next day Erdős told Selberg that he had proved the theorem; Selberg's doubts had been unfounded. In fact, Erdős had proved a slightly more powerful version of his theorem, having added the ingredient that had prevented Selberg from proving the Prime Number Theorem for himself—a proof he had told Erdős that he believed impossible. Selberg raced home and by Sunday had used Erdős's theorem to complete the elementary proof of the Prime Number Theorem. Erdős was delighted and assumed that he and Selberg would publish a joint paper on their triumphant collaboration. Perhaps Selberg would eventually have allowed Erdős to talk him into a joint publication. Before that could happen, Selberg made a quick visit to Syracuse University, where he heard rumors that would eliminate the possibility of his ever consenting to share credit with Erdős or even to discuss mathematics with him.

As Erdős always did when he had interesting mathematical news, as soon as he and Selberg had found their elementary proof of the Prime Number Theorem, Erdős immediately began sending out postcards with the news to his far-flung correspondents. Selberg in the meantime had written only to one of his brothers. Selberg was therefore surprised when, toward the end of the summer on his visit to Syracuse, he learned that the news had spread. Someone on the Syracuse faculty innocently told Selberg that Erdős had found an elementary proof to the Prime Number Theorem. According to Selberg, everyone he met attributed the proof "entirely or at least essentially to Erdős." In Ernst Straus's retelling, which became widely known, the incident became even more humiliating. According to Straus, Selberg was greeted by a breathless professor who asked: *"Have you heard the exciting news of what Erdős and some Scandinavian mathematician have done?"*

In a 1987 *Atlantic Monthly* article about Erdős, Paul Hoffmann would relate Straus's version of the story. Hoffmann went on to suggest that Selberg was so offended by the rumors that he immediately sat down and dashed off a solo paper on the proof, robbing Erdős of the credit. Hurt he most certainly was, but Selberg had never been enthusiastic about a joint paper with Erdős; his experience at Syracuse could only have confirmed for him the wisdom of separate papers. Selberg wrote a terse letter to Erdős in which he declared, "I cannot accept any agreement with a joint paper." By then Selberg had found another way to prove the Prime Number Theorem from his fundamental equation that did not depend upon Erdős's contribution. He told Erdős that he would publish this proof along with a "brief sketch of the first proof in the introduction," in which he would acknowledge Erdős's results. Erdős, he said, could write a paper of his own in which he gave the details of the formula he had derived, but should not mention the Prime Number Theorem. Erdős was furious.

Erdős wrote to Selberg, reminding him that when they had spoken about using Selberg's fundamental formula to prove the Prime Number Theorem, "you were very doubtful of success, in fact stated that you believed to be able to show that the FUND.LEMMA [the fundamental formula of Selberg's that Erdős learned from Turán] does not imply the PNT. . . . if you would have told me [all you knew] I would have finished the proof of the PNT on the spot." Selberg's fabrication had not succeeded in discouraging Erdős and now was backfiring. It is impossible to say whether Selberg, without Erdős leading the way, would have found the proof of the Prime Number Theorem. But it is clear that Erdős, taking Selberg's attempts at misdirection as truthful, was justified in believing that he had played a crucial role in solving a problem that Selberg had claimed to be impossible.

"I completely reject the idea of publishing [only the intermediate result]," Erdős continued, "and feel just as strongly as before that I am fully entitled to a joint paper." Aware that he could not realistically expect Selberg to agree to a joint paper, Erdős proposed to write a paper of his own laying out "our simplified proof, giving you of course full credit for your share (stating that you first obtained the PNT, using some of my ideas and my theorem)." To guard against inadvertent distortions of credit, Erdős wrote: "I will of course gladly submit the paper to Weyl first, if he is willing to take the trouble of seeing that I am scrupulously fair to you."

The mathematicians finally agreed that Selberg would submit his paper to the prestigious *Annals of Mathematics* while Erdős would send his to the *Bulletin of the American Mathematical Society*. To his surprise, Erdős's paper was rejected by the *Bulletin*, probably on the advice of Weyl.

In a letter in February 1948 to the paper's referee, Weyl wrote, "I had questioned whether Erdős has the right to publish things which are admittedly Selberg's.... I really think Erdős's behavior is quite unreasonable, and if I were the responsible editor I think I would not be afraid of rejecting his paper in this form." When Erdős learned of the rejection he immediately submitted the paper to the *Proceedings of the National Academy of Sciences*, which published it.

The dispute over the elementary proof of the Prime Number Theorem has been the subject of mathematical legend and speculation for fifty years. Only after Erdős's death have the records been opened for examination, and what they reveal is a story with no clear heroes or villains. Instead it is a story that highlights a major change in how mathematics is done.

Before the twentieth century, collaboration in mathematics was rare; great results from that era mostly bear a single name. Today, hyphenated theorems are common. The same trend is seen in all the sciences, and probably has a lot to do with the explosion in the size of the scientific community—most of the scientists who ever lived are alive today—and the increasing ease of travel and communication. Whatever the reason, some scientists will always prefer to work alone, to hold onto their ideas until they are perfected. Some scientists keep their brains always open, while others keep their doors ever closed. To assure the kind of isolation he required to prove Fermat's theorem, Andrew Wiles worked alone in his attic for seven years, not telling even his closest colleagues what he was up to. Perhaps he learned a lesson from the tale of Erdős and Selberg.

The stakes that mathematicians play for also ensure that battles of priority will continue. "Mathematical fame," Hardy wrote, "if you have the cash to pay for it, is one of the soundest and steadiest investments." A mathematical truth is eternal and transcends culture. The theorems of Pythagoras and Euclid are as fresh and relevant today as when they were created. "Archimedes will be remembered when Aeschylus is forgotten," Hardy wrote, "because languages die and mathematical ideas do not." As Erdős liked to say, "mathematics is the surest way to immortality." For Erdős, immortality shared was immortality nonetheless. Ultimately all mathematics is collaborative, because all mathematicians, in Newton's phrase, stand on the shoulders of giants. (Paul Winkler, a mathematician from Lucent Technologies who worked with Erdős, liked to paraphrase Newton and say, "If I can see a bit farther it is because I stand on the shoulders of Hungarians.")

In 1950 Selberg received the Fields Medal, the mathematics version of the Nobel prize, for the elementary proof of the Prime Number Theorem and the development of what is known as the Selberg sieve method. The Fields Medal is awarded every four years to up to four mathematicians, a practice of which Erdős approved: "As long as there are two or at most four Fields Medals (every four years), nobody will be offended seriously if he doesn't get it, as long as good people get it." By all accounts, being passed over by the Fields Medal did not bother Erdős. In 1952 he was the recipient of the Cole Prize, nearly as prestigious, for his work on the proof.

Though he never spoke of it publicly, Erdős was hurt by the confrontation with Selberg. Erdős never hesitated to share credit fully for discoveries and often minimized his own contribution. But in a lengthy interview with his friend László Alpár that appeared in a Hungarian journal, Erdős's feelings became clear when he remarked: "In 1948–49 my most important accomplishment was an elementary proof of the Prime Number Theorem. . . . At the same time A. Selberg, a Norwegian, but resident of the USA, obtained similar results." In this retelling, forty years later, Selberg's role has become a footnote. Even today the proof is known as the Erdős-Selberg proof in Hungary and the Selberg-Erdős proof in Princeton.

The elementary proof of the Prime Number Theorem was an overnight sensation, but it too has become almost a footnote. The proof, while beautiful, turned out not to have the revolutionary effect anticipated by Hardy. No books were cast aside, no theories rewritten. Despite the strong connection between the zeta function and the Prime Number Theorem, the elementary proof did not illuminate the mystery of the Riemann hypothesis. "In retrospect it turns out to be not such an important piece of mathematics at all," Melvyn Nathanson, a close friend of both Erdős and Selberg, says. "Some Hungarians run around saying that Selberg didn't deserve to get this Fields Medal . . . and Erdős should have gotten it, and all of that is rubbish. It really is unfair to Selberg and it's unfair to Erdős, because both of them did quite important work apart from this and either one of them could have easily gotten the Fields Medal."

The elementary proof of the Prime Number Theorem was the fulfillment of Erdős's boyhood dream as well as the cause of one of his life's bitterest episodes. According to friends, Erdős would sometimes lament that the Selberg incident had forever barred him from a position at the Institute, where Selberg would spend the rest of his career, and caused his forty years of wandering the world, jobless and homeless. That is surely an exaggeration, but it contains a grain of truth. In 1948 Erdős would leave the United States for the first time in a decade. He would return but would never again spend so much time within the borders of any country.

CHAPTER NINE

SAM AND JOE
AND UNCLE PAUL

D URING his ten years in the United States Erdős never gave up hope of someday returning to Hungary. His observations of the antics of Sam and Joe—Erdősese for the United States and the Soviet Union—caused him to become distrustful of large governments, and in his ten years in the United States Erdős took no steps toward acquiring citizenship. For all those years Erdős did not bother to try to upgrade his student visa, which caused severe administrative headaches for him when he wanted to leave the United States for a trip to Europe, including Hungary, in 1948. Eventually Erdős was able to obtain the necessary green card to allow him free reentry to the United States.

The first stop on his journey was the Netherlands, where he collaborated on problems in combinatorics, number theory, and analysis with the leading Dutch mathematicians. In Amsterdam Erdős also met Alfréd Rényi, a young mathematician whom he had known in Budapest. In the small world of the Budapest intelligentsia, Erdős got to know Rényi through his parents, who had known the Rényis for years. While at the University Erdős's parents had attended aesthetics classes taught by the philosopher and literary critic Bernát Alexander, Rényi's maternal grandfather. Rényi's father was an engineer, and his son appears to have acquired talents and interests from both sides of his family. Rényi was a brilliant student of classical Greek language and philosophy. He was also fascinated by astronomy, which led him by a natural progression to physics and ultimately to mathematics.

In 1939, after finishing high school, Rényi fell victim to the racist *Numerus Clausus* laws limiting the number of Jews admitted to the University. He labored for a half-year at the Ganz shipyard before winning competitions in Greek and mathematics that allowed him to attend the University. Rényi studied number theory, a field in which he would later become famous, with Paul Turán.

After graduation in 1944 Rényi was called up for forced labor. He somehow managed to escape from the labor camp before his group was evacuated to the West and lived in Budapest using false documents. According to Erdős, Rényi was a resistance hero who saved potential victims of the *Nyilas*—the Hungarian Nazis who were responsible for torturing and butchering tens of thousands of Jews in Budapest and on the Western marches—by boldly disguising himself in a *Nyilas* uniform. "Whenever I met him in those days," Turán would write, "I was amazed at his level-headedness and courage." In the midst of all this Rényi somehow managed to complete his Ph.D. at the University of Szeged, the second largest university in Hungary, in a town not far from Budapest. In 1946 life had quieted down enough for Rényi to travel to Leningrad, where he was able to work and study in peace for eight months. "His progress during these few months—the first time in his life that he could concentrate fully on mathematics—was truly amazing," Turán wrote. With only a rudimentary knowledge of Russian, he managed to absorb the substance of works by the top number theorists Ivan Vinogradov and Yuri Linnik; master the theory of probability, which would become the basis of his most important work; and write several pathbreaking papers. "By an effort of will he had effaced his memories of the war years and of the forced-labor camp," Turán wrote, "to center now on his work all the fiery energy of his youth and of his exceptional gifts of understanding and concentration."

When Erdős met Rényi in 1948, Rényi was no longer a promising student but a renowned mathematician. The primary basis of his fame was his spectacular progress toward the solution of one of the most notoriously difficult problems in mathematics, the Goldbach conjecture.

In 1742 a mathematician named Christian Goldbach wrote a letter to Leonhard Euler in which he speculated that every even number greater than 2 can be written as the sum of exactly two primes, for example: $24 = 19 + 5$ and $72 = 19 + 53$. On the surface the conjecture seems reasonable enough, and almost obvious. After all, every even number can be written as the sum of two *odd* numbers in lots of different ways, and every prime greater than 2 is odd. It's easy to look for a counterexample, but despite countless hours of pencil-pushing and computer time—by 1993 the numbers up to 400 million had all been checked—none have ever been found, nor does anyone seriously believe that one ever will be. Erdős liked to point out that Goldbach's conjecture had actually been anticipated about a hundred years earlier by Descartes. "I feel that the name Goldbach's conjecture should remain," Erdős reasoned, demonstrating his strong sense of justice. "First of all, Goldbach popularized it by writing to Euler. And also, Goldbach is so poor and Descartes is so rich, it would be like taking candy from a baby."

Over the past two and a half centuries, Goldbach's candy has proved to be something of a mathematical jawbreaker. The odd version of the Goldbach conjecture, which states that every odd number greater than or equal to 9 can be written as the sum of three primes, had been partially settled by Vinogradov in 1936;* Erdős would always recall the day his friends George and Esther Szekeres were married, because it was one day after he learned of Vinogradov's proof. But the even conjecture is still unsolved and, according to the admittedly unreliable instincts of mathematicians, is likely to remain

---

* Vinogradov actually proved that every sufficiently large odd number could be expressed as the sum of three primes. By "sufficiently large" Vinogradov meant the truly gigantic number $3^{3^{15}}$. Written out this number has 6,846,170 digits!

so for a very long time. But in 1947 Rényi came tantalizingly close to a proof of the conjecture when he showed that every even number can be expressed as the sum of a prime number and a number that is *almost* prime. An almost prime number is a number that has only a small number of prime factors, where the word "small" can be made mathematically precise. Rényi's result was later improved by Chen Jingrun, who was able to show that every even number could be expressed as the sum of a prime and a number that has *two* prime factors, which is as close as a number can come to being prime without actually being prime.

Rényi would become one of Erdős's most important collaborators, co-authoring thirty-two papers before his untimely death in 1970, a few weeks short of his forty-ninth birthday. Their long collaborative sessions were often fueled by endless cups of strong coffee. Caffeine is the drug of choice for most of the world's mathematicians, and coffee is the preferred delivery system. Rényi, undoubtedly wired on espresso, summed this up in a famous remark that is almost always attributed to Erdős: "A mathematician is a machine for turning coffee into theorems." A lemma is a small theorem, usually invented to help prove a more important theorem. Turán, after scornfully drinking a cup of American coffee, invented the corollary: "Weak coffee is only fit for lemmas."

When they met in Amsterdam, Erdős and Rényi worked on number theory and wrote a paper on consecutive prime numbers, but their papers would eventually range all over mathematics, reflecting their eclecticism. Both men shared a love of probability theory, which they managed to apply to a surprising range of problems that often led to real-world applications. Often their purely mathematical investigations had at least a flavor of real-world applicability. For instance, one paper they wrote considered the question of how many different flights would be required in a country with n airports of limited capacity so that no passenger need change planes more than once. The problem, which turns out to be reducible to a problem in extremal graph theory, is unlikely to arise in precisely the same form in the real world, but the kind of results they examined are relevant to actual transportation and communication networks.

The most original and far-reaching result of the Erdős–Rényi collaboration is contained in a classic series of papers with the mysterious title, "On the Evolution of Random Graphs," written in 1959. Written to satisfy the authors' purely mathematical curiosity, those papers may turn out to hold one of the keys to explaining a wide variety of real-world phenomena, including the origin of life.

In 1947, to prove a theorem concerning Ramsey numbers, Erdős came up with the strange idea of considering random graphs. As the name implies, a random graph is a graph constructed not by careful design but by chance events. Imagine that some mad civil engineer determined which cities to connect by roads by tossing a coin: heads, build a road between Albany and Boston; tails, do not. The resulting haphazard network of roads is a random graph. Erdős used random graphs in Ramsey theory to solve a generalization of the party problem—how large does a party have to be before it must contain N people who all know each other *or* N people who are strangers? As we discussed earlier, nobody knows the answer when N is 5 or more. Using random graphs,* Erdős invented an ingenious method of finding a lower bound to the necessary number of guests. Imagine, for example, that we are trying to determine how large a party must be to guarantee that it has the property that at least seven guests all know each other or seven guests are all strangers. Erdős calculated the probability that a random party with G guests *does not* have this property. If the probability is also greater than zero and less than one, then the likelihood that a randomly chosen party with G guests *does* have the property is greater than zero, since the party either does or does not have this property. For instance, if Erdős's calculation had determined that the probability that a party of two hundred guests did not have the property that seven of them were acquaintances or seven of them were strangers was .99, or 99 in 100 (this number is not the actual real probability), then the chance that the party does have that property is .01, or 1 in 100. "That means that

---

* Recall that the party problem, introduced on page 82, can be translated to a graph theory question by imagining each guest to be the vertex of a graph. Two vertices are connected by an edge if and only if the guests represented by the vertices know each other. A group of N people, all of whom know each other, is then the same as a set of N vertices, each of which is connected to all other vertices by an edge; a group of N people, all of whom are strangers, is simply a set of N vertices, none of which are connected to any of the others.

there must—not probably—there must exist graphs with that property," explains Erdős's collaborator, Joel Spencer.

The idea of using probability to prove a mathematical result that Erdős used in his 1947 paper was brand new. In the hands of Erdős, Spencer, and many other mathematicians the probabilistic method—often called the Erdős method—has become a powerful tool with which to solve previously intractable problems. "It has a magical element to it," Spencer admits. Erdős, with his probabilistic method, was a magician who proved to his audience that his hat contained a rabbit and then moved on to other tricks. The menial task of hoisting the bunny out by its ears, Erdős left as an exercise to others.

The rabbits in Erdős's hat have often proved to be solutions to real-world problems concerning things like the design of computers or information networks. Erdős's probabilistic method guarantees that solutions exist, but finding them is another matter. Knowing that a haystack contains a needle does not necessarily mean that it's practical to find it by sifting through the hay strand by strand. To assure that the search succeeds in a reasonable amount of time, some trick, like using a magnet to attract the needle, must be found. In mathematics those tricks are called algorithms—systematic procedures for solving problems, usually with the help of computers. "A very interesting theme over the last couple of decades," Spencer says, "has been called 'from Erdős to algorithms.' " Since the 1970s theoretical computer scientists and mathematicians have developed ways to convert Erdős's Platonic musings into algorithms that can solve real problems. Erdős was also a champion of graph theory and combinatorics, two branches of mathematics once considered backwaters but now essential tools of computer science. Erdős himself never touched a computer; when the Internet became an increasingly important part of mathematical culture he managed to get friends to send and receive e-mail for him, Spencer says, "so it's ironic that he had such an enormous influence on the development of theoretical computer science."

A theme that runs through much of Erdős's work is the subtle relationship between order and chaos. The clearest expression of this theme is found in his work in Ramsey theory, which demonstrates that complete disorder is impossible. In 1959 Erdős and Rényi pursued this theme by looking for order in the jumbled tangle of random graphs, structures deliberately contrived to be formless. To their surprise, even in this most random of situations, orderly structures arise spontaneously.

The situation that Erdős and Rényi analyzed is a variation of the previously described mad civil engineer scenario. Once again the engineer has the task of building roads connecting a large number of cities, say 10,000. He starts by ignoring distance and choosing two cities at random and building a road between them. When he's finished with that, the engineer chooses two more cities at random and builds another road. The engineer proceeds in this fashion, not bothering to build a second road if two cities are already connected.

At first roads will connect only a few cities. But as the engineer adds roads small, interconnected clusters of cities will form. People living in those clusters can drive to any other city in the cluster along a series of roads. Erdős and Rényi found that at first the clusters of cities are small and scattered. As the engineer adds more roads, the size of the clusters grows very slowly, and the cities within the clusters become more interconnected. Nothing much changes until the number of roads edges up to half the number of cities, when a miracle of sorts occurs. Suddenly, with the addition of surprisingly few roads, many of the the isolated clusters become interconnected and merge to form a giant cluster that includes almost all the cities.

The rapid transition from small, isolated clusters to a single giant cluster bears a striking resemblance to many natural phenomena, like the sudden freezing of water or the jamming of traffic. These phenomena, known as phase transitions, have long fascinated and baffled scientists. Erdős's and Rényi's analysis of random graphs, undertaken to satisfy purely mathematical curiosity, provided a simple model that has helped illuminate the mechanism of phase transitions. "This paper started an entire field," Erdős's disciple Joel Spencer observes, "and you look back and you can see that all the developments stem from this one idea of what happens when you throw in edges at random." Since the publication of the Erdős–Rényi paper on random graphs there have been hundreds of other papers, numerous books, and international conferences on the subject.

The practical implications of their research were not entirely lost on the pure mathematicians. In their original paper they write that the emergence of structures in random graphs "may be interesting not only from a purely mathematical point of view. In fact, the evolution of graphs may be considered as a rather simplified model of the evolution of certain communication nets (railways, roads or electric network systems, etc.) of a country or some other unit. . . . It seems plausible that by considering the random growth of more complicated structures one could obtain fairly reasonable models of complex real growth processes (e.g., the growth of a complex communication net consisting of different types of connections, and even of organic structures of living matter, etc.).".

More than forty years later this last observation has proved to be remarkably prescient. The Santa Fe Institute's Stuart Kauffman drew heavily on the evolution of random graphs to create his compelling theory on the origin of life. In Kauffman's model life emerges from the primordial stew of molecules with the same inevitability as clusters emerge from Erdős's and Rényi's model of the evolution of random graphs.

Kauffman begins by imagining a random soup of molecules, the kind of chemical mélange that scientists simulating early conditions on the earth cook up in laboratories. The soup will probably contain by chance pairs of molecules that, with the aid of a third molecule called a catalyst, can join together and form a new molecule. Every so often the new molecule will find a partner of its own, perform a similar dance with the help of another catalyst, and create yet another molecule. With luck the process might continue, the new molecule might find a mate and a suitable catalyst, and so on. The result is a long daisy chain of interactions. With even more luck this chain will loop back upon itself like a snake swallowing its tail, and will form a self-sustaining web of chemical reactions, a closed chemical economy. In other words, life.

The emergence of such a complex, self-sustaining network of chemical reactions—what Kauffman calls an autocatalytic network—depends upon a lot of dumb luck and therefore seems unlikely, to say the least. But Kauffman noticed that the network of possible chemical reactions resembles a random graph, with the molecules as nodes and the catalyzed reactions as edges. Kauffman showed that just as a random graph with relatively few edges undergoes a phase transition from disconnected to connected, a random stew of chemicals can undergo the phase transition from unrelated molecules to a living system. "The rather sudden change in the size of the largest cluster," in a random graph, Kauffman writes, "is a toy version of the phase transition that I believe led to the origin of life.... When a large enough number of reactions are catalyzed in a chemical reaction system, a vast web of catalyzed reactions will suddenly crystallize. Such a web, it turns out, is almost certainly autocatalytic—almost certainly self-sustaining, alive." The mathematics of the evolution of random graphs shows how the seemingly unlikely emergence of autocatalytic systems may in fact be inevitable. Given a large enough diversity of molecules "a self-reproducing chemical system . . . springs into existence."

After spending several months in Amsterdam, in December 1948 Erdős returned to Budapest for the first time in over a decade on a special visa arranged for him by the Ministry of Public Education that would allow him to leave again for the West. "At that time this was exceptional treatment," Erdős would say. He was gladdened to find that many of his closest friends had survived the Holocaust. Erdős had seen Turán in the United States, of course, but was happy to be reunited with some of the other young mathematicians he used to meet in the Budapest City Park at the Statue of Anonymous: Tibor Gallai, László Alpár—between his stints as a political prisoner—and others. But his greatest joy was to be reunited with his *Anyuka*. "I found my mother in our old home," he later reported happily. "She was energetic and seemed healthy." The joy of his reunions was tempered by the enormous toll exacted by the Nazis among his family and friends and by the death of his father. Of his half-dozen or so closest relatives only his mother and one aunt survived.

Erdős was also deeply troubled by the worsening political situation. Like most of his childhood friends, Erdős was a liberal. For him that term seems to have meant mostly a strong belief in equality and an abiding concern for humanity and the needs of individuals. Those principles were coupled with a deep suspicion of political power, which throughout his life put him at odds with both Sam and Joe. Toward the beginning of 1949 Joe had begun a series of horrifying public show trials. If Erdős had entertained the idea of returning to Hungary for good, he quickly changed his mind. "Because of the new turn in the political situation I felt it was advisable to stay away from Hungary." After a three-month visit to Budapest Erdős packed his few belongings and resumed his travels. Erdős's worst fears were confirmed when a few months later his friend Alpár was once again arrested, this time for his association with the Minister of the Interior, László Rajk, who had been executed as a spy after the first of the show trials. Alpár labored for four brutal years in a copper mine before being freed in the thaw following Stalin's death in 1953.

For the next several years Erdős divided his time between the United States and Great Britain, living off loans and stipends from short lectureships. In 1950 a large mathematical conference attended by mathematicians from all the Communist countries took place in Budapest. Fearing that he would not be permitted to leave Hungary after his visit, Erdős's Hungarian friends did not invite him, even though Lipót Fejer, Erdős's teacher and friend, was being honored. That same year the Hungarian Institute of Mathematics was founded along Soviet lines. With Rényi as director—a position he held until his death in 1970—the Institute would become a leading mathematical center and Erdős's most important refuge in Eastern Europe.

In 1953 it looked as though the United States would become Erdős's permanent residence. Arnold Ross, chairman of the mathematics department at the University of Notre Dame in South Bend, Indiana, invited Erdős to spend a year on extraordinarily generous terms. Ross had arranged things so Erdős had to teach only one advanced course and provided him with an assistant to take over whenever he was seized by the urge to travel.

Erdős was an avowed atheist, and his friends at Notre Dame enjoyed teasing him about his working at a Roman Catholic university. "He said in all seriousness that he liked being there very much," Melvin Henriksen, a colleague from those days, recalled, "and especially enjoyed discussions with the [priests]." Only one thing bothered him. "There are too many plus signs," he irreverently remarked.

Henriksen likes to recall how his only paper with Erdős came about. Henriksen and Leonard Gillman were working on a problem in topology, a field in which Erdős was uninterested. Along the way they ran into a set-theory problem, a field in which Erdős was an acknowledged master. They presented Erdős with their problem, which he quickly solved, earning himself another co-authorship. When the two topologists tried to explain to Erdős the motivation for their questions, his eyes glazed over. "I have often said that Erdős never understood our paper; all he did was the hard part," said Henriksen. The paper became one of the pioneering works in a field known as nonstandard analysis, and with alphabetic injustice is often credited to Erdős et al.

After a year at Notre Dame, Ross made an offer Erdős could seemingly not refuse, to continue his appointment indefinitely on the same comfortable terms. According to Henriksen, Erdős thanked Ross politely and refused. His friends thought he was crazy. "Paul, how much longer can you keep up a life of being a traveling mathematician?" they asked, little suspecting that the answer would be more than forty years. Nevertheless, whether or not Erdős had decided to remain at Notre Dame would soon prove immaterial. Once more a man named Joe would cause Erdős to change his plans.

Senator Joseph McCarthy's hysterical campaign to rid the United States of the Red Menace was reaching its vitriolic peak. Erdős received his first unpleasant taste of McCarthyism on July 6, 1953. While visiting a friend in Los Angeles, Erdős asked to use the phone to call his mother in Budapest and wish her a happy seventy-third birthday. Erdős's friends were accustomed to footing his long-distance bills and usually did so with no more than an admonition to keep it short. But this time the friend refused to let him make the call, not out of frugality but out of fear. He did not want to have a call to a Communist country appearing on his phone bill.

That refusal may have been cowardly, but it was not altogether foolish. Ever since the passage of the McCarran Internal Security Act in 1950, scientists had reason to feel paranoid. Foreign scientists wishing to visit the United States had to submit to a humiliating investigation in order to obtain a visa. An editorial in *The Saturday Evening Post* in 1954 defended the Act, claiming that no scientist would be barred unless "he turned out to be a really bad egg." Evidently, in the eyes of the United States government, Paul Dirac, the brilliant British Nobel prize–winning physicist, fitted this description. The astronomer Otto Struve expressed his outrage that such a misguided policy should deny American scientists the benefit of a visit from Dirac. At any rate, he concluded, if Dirac is a bad egg, "we should not be too reluctant to add a small dose of that kind of egg to our domestic diet."

Many scientists were reluctant to apply for a visa, fearing that a refusal would result in their being labeled as "red" or "pink" by their own governments. American scientific organizations began to move their meetings overseas to make it easier for foreigners to attend. The American Psychological Association had hoped to hold the 1954 International Congress of Psychology in New York City but finally decided to hold it in Montreal, "because of the delays and embarrassments which foreign scientists experience in attempting to obtain even temporary admission to this country." Unfortunately, holding conferences overseas raised problems for foreign scientists who, like Erdős, lived in United States.

In 1954 Erdős wanted to attend the International Congress of Mathematicians in Amsterdam, a large and important get-together held every four years. Erdős was not a U.S. citizen, so if he wanted to return to the United States he had to obtain a reentry visa. The last couple of times he had left the United States the process of obtaining a reentry visa was largely a matter of writing a few letters and managing some bureaucratic wrangling. This time, thanks to Joe McCarthy and the McCarran Act, the Immigration and Naturalization Service wanted to have a chat.

The INS sent an agent from Detroit to Notre Dame to interview Erdős in his office. Erdős appreciated the courtesy but was offended by the interview that followed. The agent informed him that the United States had been keeping close tabs on Erdős's activities. For example, there was the time he had been arrested for prowling around a radar installation on Long Island, with two foreigners no less. And there was his correspondence with a Chinese number theorist named Lo Ken Hua, who had returned to Communist China in 1949. Little matter that, as Henriksen pointed out, "a typical Erdős letter would have begun: Dear Hua, Let p be an odd prime . . ." Erdős also wrote to his mother, who had become a Communist Party member in order to keep her job at the Hungarian Academy of Sciences. Guilt by association was the order of the day. And many of Erdős's associations did not look good.

Erdős answered all the agent's questions with his usual guileless honesty. Would Erdős return to Hungary if he were convinced that the Hungarian government would let him freely leave? "Of course," Erdős replied, "my mother lives there and I have many friends there." What do you think of Karl Marx? the examiner asked. Erdős admitted that he had only read *The Communist Manifesto*, so "I'm not competent to judge. But no doubt he was a great man."

Erdős, according to his friends, had always been a leftist, but never a Communist. As a matter of survival Erdős had become deeply interested in politics, and when he was not proving and conjecturing he enjoyed "Sam-ing and Joe-ing," that is, talking about politics. But Erdős himself had never belonged to a political party. "He had a fierce belief in the freedom of individuals as long as they did no harm to anyone else," Henriksen explains. Any country that did not share this principle Erdős dismissed as imperialist and, with a mathematician's strong sense of symbols, designated by a lowercase name. The United States became samland and the Soviet Union joedom. Erdős joked about an organization that he called the fbu, a cross between the FBI and the Soviet predecessor to the KGB, the OGPU. Erdős could not help being a Hungarian citizen, but in his years of traveling he had never tried to become a citizen of any lowercase country.

Erdős's belief in the possible greatness of Marx, coupled with his refusal to denounce his family and friends, was enough for the government to decide he was a fellow traveler who constituted a threat to the United States. His request for a reentry visa was denied. Erdős hired a lawyer, wrote letters, and sought the help of friends, but the INS would not budge. Erdős had a green card and was welcome to stay in the United States. But if he left the country his green card would be confiscated and he would not be permitted to return. Naturally, Erdős left. "Since I don't let Sam and Joe tell me where I am traveling, I chose freedom," he would explain. "I still feel I acted in the best traditions of America: You don't let the government push you around."

The night before he left for Amsterdam Erdős had dinner with his friend Harold Shapiro. Like all of Erdős's American friends, Shapiro had been trying to persuade Erdős to stay until the difficulties with the government could be worked out. "I should knock you on the head and tie you up to stop you from leaving!" Shapiro shouted at him. "Okay, then tie me up," Erdős said. Nothing short of bondage would have kept him from leaving.

Confident that he could obtain Dutch and English visas, Erdős attended the conference in Amsterdam. To his dismay, the Dutch issued him a visa that was good for only a few months, and the British would not let him in at all. Erdős took refuge in Israel, whose constitutional Law of Return recognized the right of all Jews, even nonbelievers, to immigrate and acquire citizenship. Forced by circumstances, Erdős reluctantly accepted Israeli citizenship, although he considered Israel to be just another lowercase country. He would eventually become a "permanent visiting professor" at the Technion in Haifa, receiving a small salary for the months when he chose to visit. In gratitude to Israel, Erdős would donate most of the $50,000 he received for the Wolf Prize in 1984 to endow a chair at the Technion in his mother's memory.

Erdős's friends in the United States wrote letters and signed petitions on his behalf. Apparently none of their appeals did any good. "In subsequent years, his requests for a visitor's visa to attend conferences in the U.S. were repeatedly turned down," László Babai wrote in a memoir published in honor of Erdős's eightieth birthday. Sam's resolve wavered only once. On March 25, 1959, William Pierce, a mathematician at Syracuse University who had helped spearhead a letter-writing campaign on Erdős's behalf received two telegrams:

STATE DEPARTMENT INFORMED MY OFFICE THIS AFTERNOON THAT
TELEGRAM HAD BEEN SENT TO CONSUL IN BUDAPEST INSTRUCTING HIM
TO ISSUE VISTORS VISA TO ERDOS. KNOW YOU WILL BE PLEASED TO
HEAR THIS NEWS =

    WILLIAM H MEYER MEMBER OF CONGRESS

WAIVER FOR DR. PAUL ERDOS APPROVED TODAY. IT WAS A PLEASURE TO
ASSIST =

    HUBERT H HUMPHREY

Within a month the chairman of the Notre Dame math department, Arnold Ross, would write to the United States Consul in Budapest on behalf of "our very eccentric, very gifted, and very useful colleague" to help expedite Erdős's visa. Ross had invited Erdős to spend the 1959–60 academic year at Notre Dame to "help with our technical program of training as well as with our special program of the training of secondary school teachers. It is not generally realized that in addition to his talents as a research man, Professor Erdős possesses great talent as a teacher."

Once again, Notre Dame was to be denied the benefits of Erdős's presence. The waiver obtained by Humphrey allowed Erdős only a brief visit, enough time to attend an American Mathematical Society meeting in Boulder, Colorado, and to present a few colloquia. Henriksen remembers picking Erdős up at the airport when he visited Purdue and being astonished to see him burdened with a small suitcase. "For many years he traveled only with a small leather briefcase containing a change of socks and underwear, a wash-and-wear shirt, and some papers and preprints."

After this visit Erdős resumed his exile. Babai writes that around 1962 Erdős wrote to friends that apparently "U.S. foreign policy is adamant on two points: non-admission of Red China to the U.N. and non-admission of Paul Erdős to the U.S."

Canada evidently did not share its southern neighbor's fears. Erdős was a frequent visitor at Canadian universities, and his friends from the United States would often travel to meet him there, like loyalists visiting an exiled ruler. But Erdős rarely stayed in one place for very long. Erdős was a leading expert on the mathematics of

Brownian motion, the random jiggling motion of microscopic particles. He was also an excellent approximation of a Brownian particle himself, moving unpredictably from place to place. One friend tells of being summoned to meet Erdős at the University of British Columbia only to arrive and find that Erdős had moved on. The friend continued the pursuit and, after a comical series of missed connections across Canada, finally caught up with Erdős.

Erdős's plight began to attract the attention of the press. The headline in one Canadian newspaper trumpeted Erdős's claim to have been "Barred by 'U.S. Iron Curtain.' " In another article, which told of thirty U.S. mathematicians who held an informal Erdős conference, the University of Michigan math professor George Piranian speculated that Erdős "aroused the suspicions of some minor official whose mother had been frightened by a senator from Wisconsin." The recently launched Soviet Sputnik satellite was frightening the United States into a new awareness of the importance of education in math and science. "We may hope that [Sputnik] will create a greater measure of enlightened self-interest on the part of our government," Piranian sniffed. "It is a foolish child that spites his face by cutting off his nose."

In a few years the anti-Communist paranoia began to abate, and the efforts of Erdős's friends began to get results. The INS decided to review the case against Erdős once more. According to Babai, they were still worried that Erdős had joined "proscribed organizations." With the help of his old Cambridge University friend Harold Davenport, Erdős composed a reply in which he explained that the only organizations that he had joined were the American and British Civil Liberties Unions. After some additional clarification of his connection to the Hungarian Academy of Sciences, to which he had been elected in 1956, Erdős finally received permission to visit the United States in 1963. From then on he never had visa problems again. At his lectures Erdős liked to claim that "Sam finally admitted me because he thinks I am too old and decrepit now to overthrow him."

D URING his years of exile from the United States Erdős began to spend more time in Hungary. The Hungarians were extremely proud of Erdős's international fame and had issued him a consular passport that allowed him to come and go as he pleased. Erdős took seriously his election to the Hungarian Academy of Sciences, arranging to return to Budapest to attend meetings of the General Assembly. But he did not take the honor too seriously. When friends were elected to join him in the Academy he would congratulate them by saying, "I am glad that you became a demigod!"

Erdős happily collaborated with such old friends as Turán, Rényi, and others but actively cultivated the younger generation. On his first visit to Hungary after the war in 1948 his friend Gallai introduced him to a brilliant young high school student named Vera Sós. Between that visit and Erdős's next, eight years later in 1956, Sós had married Turán, borne her first child, completed her degree, and embarked on a brilliant career as a mathematician. She would become one of Erdős's most important collaborators and one of his closest friends and fiercest supporters.

Erdős was constantly on the lookout for new talent. On the same trip in 1956 Erdős also paid a visit to the University of Szeged, where he met a young graduate student named András Hajnal. Like many of the younger generation, Hajnal had heard of Erdős but, with travel restricted, had not met him. Hajnal was a student of Erdős's old friend Kálmár, who years ago had rewritten Erdős's first paper on the Chebychev theorem. Kálmár introduced Hajnal as a "promising young student" studying set theory and then disappeared, leaving them "sitting in two enormous armchairs facing each other over a coffee table."

At first Hajnal was nervous. "I felt very honored, and a little embarrassed, to be left alone with this famous man," Hajnal recalls. "I did not know then that he met most of his young collaborators in a similar way." Erdős asked Hajnal about his doctoral dissertation, which concerned a topic on the borderline of set theory, a subject that interested Erdős, and logic, which did not. "He had no feeling for logic, he believed in absolute truths. So this relativism—it might be true, it might not be true—seemed to bother him," Hajnal explains, referring to the strange indeterminacy unleashed by Gödel and his followers.

Hajnal was proud of his work and began to explain it enthusiastically. After listening for a while Erdős interrupted to ask, "Are you interested in real mathematics?" The question was put so politely that Hajnal did not take offense. "It was quite obvious," says Hajnal, "that he did not mean any harm."

It turned out that Hajnal was interested in "real mathematics" too. Hajnal brought up another problem, which to his relief Erdős liked. The conversation, which had been formal and polite, suddenly became lively and excited. Before long they had proved some lemmas and made a few conjectures. In the midst of that creative outpouring Erdős recalled that there was something else he had planned to do on his trip to Szeged. Near the mathematics building was what Hajnal recalls as a "rather ugly" cathedral built in the 1930s sporting two tall towers. Wherever he traveled Erdős insisted on climbing the tallest structure even when it promised to afford only a dull view. "I had by then lived for two years in Szeged and I had never had the slightest difficulty in resisting any pressure to visit the tower, especially since the surrounding countryside is absolutely flat," Hajnal says. But he could not say no to Erdős and soon found himself making the three-hundred-step climb. "It was interesting, because at that time he always felt dizzy," Hajnal recalled. "He was complaining that he was dizzy and we were very worried that the old man [Erdős was forty-three, Hajnal twenty-five] was going to fall down." But Erdős remained upright and when not complaining formulated more results and conjectures. They continued working over dinner that night at Kálmár's and parted as old friends, with their first paper half completed.

In the early years, when Erdős visited Budapest he would stay with his mother in her apartment. Hajnal stopped by frequently, but before getting down to mathematics he always had to perform two small duties. He had to solve the crossword puzzle for *Anyuka* and then chess problems for Erdős. Although he could solve the most difficult and abstruse mathematical problems while climbing a tower, out of breath and dizzy, Erdős was frequently stumped by simple chess problems. "He looked at them, got impatient, and wanted to do mathematics," Hajnal explains. But not before Hajnal showed him how to force checkmate in four moves.

When they finally got down to mathematics, Erdős could go on working almost indefinitely. In self-defense, Hajnal made a rule: no math after seven in the evening. "I just absolutely refused to work when I was tired. I would play chess with him, but that didn't satisfy him," Hajnal says. Between chess games, while setting up the pieces, Erdős would try to turn the conversation to mathematics. Hajnal would firmly say, "No, Paul, I am tired." Shortly afterward Erdős would go to his bedroom, where he wrote in his mathematical journals until late at night. Throughout his life Erdős kept a meticulous journal in which he recorded his mathematical thoughts, and the proofs and conjectures he had discussed with the many mathematicians he met in the course of a day. Erdős's remarkable ability to take up conversations dropped months earlier, seemingly in midsentence, probably was bolstered considerably by those notebooks.

Membership in the Hungarian Academy of Sciences had its privileges. The Academy maintains small resorts to allow its members comfortable natural settings in which to relax and work in peace. Two or three times a year Erdős and his mother liked to stay at Mátraháza, an Academy resort located in the wooded hills of the Mátra mountains. At Mátraháza Erdős was usually joined by Turán and his wife, Vera Sós; Rényi and his wife, Catherine, who was also a mathematician; Hajnal; and others. Most of the other Academicians at Mátraháza were old scholars and writers who ate their meals in silence and cast envious glances at the large table where Erdős and his friends noisily swapped stories, ideas, and jokes.

Erdős enjoyed meeting the other Academy members and their guests, who often included the leading lights of Hungarian arts and sciences. His irreverence toward those distinguished guests was a source of constant entertainment for his friends. After being introduced to a well-known opera singer Erdős asked, "So, where are you shouting?" When introduced to a famous poet, a man whose name was known to every schoolchild in Hungary, Erdős innocently inquired, "And what is it that you do?" After the poet told him, Erdős asked, "Can you make a living at this?" Given Erdős's status as chronically unemployed mathematician, this was indeed a case of the pot calling the kettle black.

Vera Sós's nephew Janos Pach, when he was a child, often accompanied his parents on visits to Mátraháza. Pach fondly remembers the time he spent in the mountains with Erdős and his mathematical friends as a "golden age." Pach frequently accompanied "the three Pauls"—Erdős, Turán, and his father, Paul Pach, a noted historian—on long treks through the surrounding mountains. In hiking, as in mathematics, Turán's motto was "Don't stray far from the unbeaten path," which made for long, interesting journeys and late lunches. The Pauls also shared "the same irresistible adolescent impulse to climb each and every peak they saw," Pach recalls. Even in his final years, after he had become in reality the "old and feeble" man he had always claimed to be, whenever Erdős visited Mátraháza he insisted on mounting the steep and rickety steel ladder that led to the top of a nearby observation tower.

But most of all Pach recalls how he enjoyed watching Erdős and Turán as they worked, even though the mathematics was beyond him. They were often joined by Sós and Rényi. Pach remembers Erdős skipping and jumping as ideas poured forth from him. His mind was "incredibly quick," and he would attempt to speak as rapidly as he thought, which made him difficult to follow. Turán would grow annoyed and chastise Erdős for speaking "nonsense." Sós, who was more patient at piecing together the fragments of Erdős's insights, would then calmly smooth things over. Rényi's witty asides, "which even an epsilon could appreciate, made it even more fun to watch them."

Pach sat to the side as unobtrusively as possible until the grownups got up to take a break. He then crept stealthily across the now quiet room to the abandonded work table. Pach gazed in awe at the scattered pages, dense with the arcane heiroglyphs of higher mathematics, the focus of all that noisy, serious fun. "I was astonished when I first saw the end product of their work: strange letters, numbers, signs, arrows, scribble-scrabble. . . . I had no doubt that the Laws of the Universe were written in this mysterious language. Otherwise, how could mathematical problems spark such enthusiasm in these brilliant and famous people!" Pach decided that he too would like someday to read and write that mysterious language. With inspiration and encouragement from Erdős, Pach became a mathematician and another of Erdős's army of collaborators.

The only requirement to join Erdős's army was mathematical talent. Age, in particular, was never a barrier. Whenever Erdős was in Hungary he gave eagerly attended talks at the Young Mathematicians' Club. Béla Bollobás (who insists he was not a prodigy) vividly recalls attending one such meeting when he was fourteen, in 1958. "I was totally fascinated," he says. Erdős had a talent for pitching his talks at a level understandable by his young audience. He posed problems in combinatorics, geometry, and number theory that were interesting, were easy to understand, and often had the additional allure of being unsolved. "These were problems that needed no other mathematical background," Bollobás explains. All they required was "independence of thought and ingenuity."

A few months after that talk Erdős heard about Bollobás, who had won all the math competitions for his age. As was his custom with promising epsilons, Erdős invited Bollobás to join him and his mother for lunch at a fancy Budapest hotel. Erdős and Bollobás talked mathematics while *Annus néni*—Aunt Anna, as Erdős's mother was called by all the epsilons—proudly listened. When Erdős was traveling he corresponded with Bollobás, and the two met regularly when Erdős was in Budapest. Bollobás wrote his first paper with Erdős when he was seventeen. "It was a tiny little thing," Bollobás recalls, a small problem that Erdős proposed that was precisely suited to Bollobás's talent and the current level of his skills. "He always knew which problems were good for whom," Bollobás says. He would often say, "for different horses you need different courses." Bollobás would go on to write major monographs in fields pioneered by Erdős, extremal graph theory and random graphs. When Erdős turned sixty-five, Bollobás organized a conference in his honor at Cambridge University and would continue doing so every five years.

Of the many prodigies Erdős nurtured, the one he spoke of with the most affection and regret was Lajos Pósa. When Erdős visited Hungary in 1959, he learned that "there is a little boy whose mother is a mathematician and who knows all that there is to be known in high school." Erdős was immediately interested and arranged to have lunch with the eleven-year-old wonder and Rózsa Péter, a mathematician with whom Pósa studied.

As Pósa was eating his soup, Erdős gave him a problem: "Prove that if you choose $n+1$ integers from 1 to $2n$, at least two of them are relatively prime." Two integers are relatively prime if they have no common factor greater than one. So, for example, 7 and 15 are relatively prime, but 10 and 25 are not, since they share a common factor of 5. To understand Erdős's question it helps to choose a specific value of n—say 5. In that case, $2n$ is 10 and $n+1$ is 6. According to the little theorem that Erdős asked Pósa to prove, if you chose six different integers from 1 to 10, then at least two of them must be relatively prime. This is not the case if you were to choose only five different integers, because in that case you could choose the even numbers 2, 4, 6, 8, and 10. Since every even number is a multiple of 2, no two even numbers are relatively prime.

Pósa froze for an instant, his brimming spoon poised in space, and uttered the three-word proof: "Two are consecutive." In an instant, Pósa had realized that when you choose more than half of the numbers from 1 to $2n$, two of them *must* be consecutive,* and consecutive numbers are *always* relatively prime. When Erdős discovered this simple result some years earlier it had taken him ten minutes to find the proof. "Needless to say," Erdős wrote, "I was very much impressed. . . . I think that this is on the same level as Gauss," who, recall, as a child had instantly summed the integers from 1 to 100. When Erdős told listeners about Pósa's precocious feat in lectures, he liked to quote a Canadian mathematician who said, "For this occasion champagne would have been more appropriate than soup."

From then on Erdős worked closely with Pósa, sending him many letters with problems when he was traveling and meeting with him when in Budapest. When Pósa was thirteen Erdős explained to him the infinite version of Ramsey's theorem. "It took about fifteen minutes until Pósa understood it, and then he went home, thought about

---

* To help see this, imagine a row of ten baskets. You are given six balls and told to distribute them one to a basket in such a way that no two are in adjacent baskets. Everything goes fine for the first five balls, which you place in alternating baskets, but then the sixth must be placed between two filled baskets.

it all evening, and before going to sleep he had a proof." By the time Pósa was fourteen, Erdős recalled, "you could talk to him as a grown mathematician." Erdős would phone him to discuss mathematics and found that if the problem did not presuppose a lot of sophisticated mathematics that Pósa had not had the time to learn, "if was very likely that he had some relevant and intelligent comment." Pósa's first joint paper with Erdős was published when Pósa was fourteen. Before long Pósa was publishing solo papers containing significant original results. Strangely, Pósa never really caught on to calculus, and Erdős failed to interest him in geometry. "He always liked to do only what he was really interested in, but at that he was extremely good," Erdős said with a mixture of pride and exasperation.

Taking seriously Rényi's dictum that a mathematician is a machine for turning coffee into theorems, Erdős gave the fourteen-year-old Pósa cups of the strong institute brew to drink. When Erdős's mother heard of that she scolded her son for his irresponsible behavior. "I answered that Pósa could have said: 'Madam, I do a mathematician's work and drink a mathematician's drink,'" Erdős said, adapting a line he had heard uttered years earlier by a young whisky-drinking cowpoke in a Western.

When Pósa began the ninth grade he started at the Fazekas High School, which had just initiated a special program for gifted mathematics students. The school boasted an astonishing number of talented young mathematicians, the best of whom—including László Lovász and József Pelikán—Pósa would introduce to Erdős. In addition, Erdős corresponded with a young prodigy from Szeged named Attila Máte, who would visit Budapest a few times a year to attend the Young Mathematicians' Club meetings. On one visit he phoned Erdős's apartment and when *Anyuka* answered announced, "I am the epsilon from Szeged!"

Erdős tried to teach his epsilons more than mathematics. Once Lovász and Pósa asked him why there were so few female mathematicians. Erdős, who counted many women among his collaborators, explained that the problem was not one of native ability. "I told them: Suppose the slave children (boys) would be brought up with the idea that if they are very clever the bosses (girls) will not like them—

would there be then many boys who do mathematics?" The young slaves thought it over and finally said, "Well, perhaps not so many."

While attending the University, Pósa, who "liked to do only what he was really interested in," discovered that he was more interested in teaching than in mathematics. To Erdős's disappointment Pósa stopped doing original mathematics and became a teacher. "He doesn't even like to do it at the university, he does it at the high school," Erdős complained, like a proud but baffled parent whose brilliant child, after graduating with honors from Harvard Law School, chooses to become a public defender. "He is very good at it," Erdős allowed.

Nevertheless, when Erdős spoke of Pósa in subsequent years, he would shake his head and say, "It's a pity he died at such a young age," although Pósa was alive and well.

CHAPTER TEN

SIX DEGREES OF
COLLABORATION

HE IS A GENIUS, A PHILOSOPHER, AN ABSTRACT THINKER. HE HAS A BRAIN OF THE FIRST ORDER. HE SITS MOTIONLESS, LIKE A SPIDER IN THE CENTER OF ITS WEB, BUT THAT WEB HAS A THOUSAND RADIATIONS, AND HE KNOWS WELL EVERY QUIVER OF EACH OF THEM.

Sherlock Holmes describing Dr. Moriarty in The Final Problem
(Sir Arthur Conan Doyle)

B Y the late 1950s Erdős's life had begun to resemble a travel montage in an old Hollywood movie, a series of overlapping dissolves of trains and planes and suitcases plastered with stickers from exotic countries. In 1960, for example, Erdős moved rapidly from Budapest to Moscow, then to Leningrad, back to Moscow and on to Beijing by way of Irkutsk and Ulan Bator. After three weeks in Beijing catching up with old friends—Chao Ko, whom he had met in Manchester, and Lo Ken Hua, whose correspondence had helped label Erdős as a Communist—Erdős caught a flight to Shanghai, then boarded a train to Hangchow. Another flight took him to Canton, from whence another train, this time to Hong Kong; then he flew to Singapore and finally on to Australia to visit George and Esther Szekeres. And that wasn't even a particularly busy year. As Richard Bellman would write: "One never knew where Erdős was, not even the country. However, one could be sure that during the year Erdős was everywhere. He was the nearest thing to an ergodic particle [one that eventually visits all physically possible states] that a human could be." A friend of Erdős once bumped into him unexpectedly on the street and asked, "Paul, are you here or are you somewhere else?"

Far from exhausting him, the constant motion from place to place, meeting scores of mathematicians and hearing hundreds of new theorems and conjectures, inspired Erdős to even greater productivity. "Wherever I went there were circles of young and old mathematicians to whom I proposed interesting problems, whose research I was able to get involved in, and with whom I was able to work," Erdős said. "In addition, I brought with me numerous unsolved problems which I heard somwhere else. This form of communication was faster and more effective than letters."

Not that Erdős was a reluctant correspondent. Every year he wrote thousands of letters and postcards in his untidy, loopy, but legible scrawl, answering questions from world-famous mathematicians and unknown students with equal celerity and courtesy. His memory for names, phone numbers, and obscure mathematical references was legendary. Mathematicians would frequently tell him about the arcane problem that they had been working on, and Erdős would close his mind for a moment, riffle through his mental card file, and come up with a precise reference to a relevant paper.

His ability to connect names and faces was not nearly as perfect. Many mathematicians share Allen Schwenk's experience: "For the first six or eight years of our acquaintanceship, Paul always greeted me the same way: 'Hello, where are you now?' Since I remained at the Naval Academy for that entire time, I thought this question was strange, until I realized that while he recognized my face as a mathematician, he couldn't place my name. The question gave him the vital clue he needed to identify me. Later, when my identity was secure, the greeting changed to 'Hello, how are your boss and epsilons?'" On another occasion, Erdős met a mathematician and asked him where he was from. "Vancouver," the mathematician replied. "Oh, then you must know my good friend Elliot Mendelson," Erdős said. The mathematician replied, "I *am* your good friend Elliot Mendelson."

At math conferences Erdős worked the lobby like a chess grandmaster playing a simultaneous exhibition. He would move from group to group, listen for a few moments to the problem under discussion, make a suggestion, and move on to the next group. The questions being hashed out were often in completely different fields of mathematics, demanding different styles of thought, but Erdős would switch almost instantly. When he returned, whether in a few minutes, months, or years, he had the sometimes disconcerting ability of taking up the argument exactly where it had been broken off. In the language of computer science, Erdős's brain was multithreading and multitasking.

As Richard Rado would write, "Wherever [Erdős] pays a visit he leaves behind him a visible paper trail: *Ex ungue leonem* [From his claw one can tell a lion]. On whatever object he casts his look theorems spring up like mushrooms." With increased travel Erdős became more prolific than ever, and the number of his collaborators grew from a crowd to a mob. It began to seem as if everyone had either written a paper with Erdős or at least written a paper with someone who had. Erdős resembled Conan Doyle's Professor Moriarty, except that he was never motionless. The web he occupied was a tangled skein of collaborations.

Someone noticed that the web of Erdős's collaborations could be precisely described by a graph, a mathematical object dear to Erdős. From that observation the concept of the Erdős Number was born. Everyone who has written a paper with Erdős is said to have an Erdős Number of 1. Anyone who writes a paper with someone with an Erdős Number of 1 receives an Erdős number of 2, and so on. If no chain of collaboration can be found connecting a mathematician to Erdős, he is said to have an infinite Erdős Number, which is a sign of either independence or nonentity. Virtually any mathematician and probably most scientists of any stripe who have written a collaborative paper have a finite Erdős Number. Einstein's Erdős Number, for example, is an impressively low 2 (he wrote a paper with Ernst Straus, who wrote one with Erdős). Swapping Erdős Numbers and Erdős Number lore is a popular party game among mathematicians. At a gathering of mathematicians a good icebreaker is, "What is your Erdős Number?"

The Erdős Number game is similar to the party game involving the actor Kevin Bacon. Film buffs compute the Bacon number of actors and actresses by assigning a Bacon number of 1 to anyone who has appeared in a movie with Bacon, 2 to anyone who has appeared in a movie with someone who has a Bacon number 1, and on and on. Depending upon how one defines "movie" (purists require that a movie have a theatrical release to qualify), Erdős has a respectable Bacon number of 4. A mathematician and sometime actor named Gene Patterson appeared briefly in the 1993 documentary about Erdős, *N is a Number*. Patterson also had a role in *Box of Moonlight* with John Turturro, who was in *The Color of Money* with Tom Cruise, who appeared in *A Few Good Men* with Kevin Bacon. With a little more stretching of definitions, Bacon can be said to have a combined Bacon–Erdős Number. A mathematician named Benedict Gross has a Bacon number of 2 for having worked as the mathematics consultant in the Jill Clayburgh film *My Turn* and an Erdős Number of 3. Thus do the worlds of Hollywood and higher mathematics overlap.

Even baseball can be linked to mathematics by way of Erdős. In 1974, during the weeks before Hank Aaron broke Babe Ruth's record of 714 home runs, the numbers 714 and 715 were on everyone's lips. Carl Pomerance, a number theorist and baseball fan from the University of Georgia, wondered whether those numbers held any interest beyond the ballpark. Discovering curious facts about numbers encountered in everyday life is a popular diversion among number theorists. The most famous instance concerns a visit by G. H. Hardy to Ramanujan, who was in the hospital. On the taxi ride to the hospital Hardy had made a note of the number of the taxi, 1729. As he entered Ramanujan's room he remarked that the taxi had "a rather dull number" and hoped that this wasn't a bad sign. Ramanujan asked the number and immediately replied, "No, Hardy, it's a very interesting number. It is the smallest number that is the sum of two cubes in two different ways." In other words, both $10^3 + 9^3$ and $12^3 + 1^3$ are equal to 1729, and no smaller number has the same property. "How did Ramanujan know?" his biographer, Robert Kanigel, asks. "It was no sudden insight. Years before, he had observed this little arithmetic morsel, recorded it in his notebook and, with that easy intimacy with numbers that was his trademark, remembered it."

After becoming intimate with the numbers 714 and 715, Pomerance soon discovered their interesting secret.* The sum of the prime factors of 714 is exactly equal to the sum of the prime factors of 715. That is,

$$714 = 2 \times 3 \times 7 \times 17$$
$$715 = 5 \times 11 \times 13$$
$$2 + 3 + 7 + 17 = 5 + 11 + 13 = 29.$$

Pomerance quickly decided to call numbers with this property Ruth–Aaron pairs, and he published a paper in an obscure journal containing a speculation on how frequently such pairs occur. The paper, which was written with Carol Nelson and David Penney, was more or less a mathematical joke, but for some reason it caught Erdős's eye. Erdős saw how to prove Pomerance's conjecture on the density, or frequency, of Ruth–Aaron pairs, and decided to give Pomerance a call.

Pomerance was astonished that Erdős had even seen his paper and absolutely flabbergasted that the famous mathematician phoned. "I was well-known to my calculus students that semester, but that was about it," he recalled. The result of the call, naturally, was— aside from some good mathematics—that Pomerance can now boast that he has the Erdős Number 1. Pomerance would go on to write more than twenty papers with Erdős.

In 1995 the University of Georgia conferred honorary degrees on Paul Erdős and Hank Aaron. Erdős invited Pomerance and his wife to the reception, where Aaron was signing baseballs. Pomerance cor-

* Actually the numbers have several "interesting properties" the most dubious of which is that $714 + 715 = 1429$. This rearrangement of the digits of the year that Columbus discovered America is a "backwards-forwards-sideways prime." That is, 9241, 1249, 9421, and 4219 are all primes.

nered Aaron and tried to tell him about prime factors, Ruth–Aaron pairs, and the rest. The baseball great "was somewhat bemused," Pomerance said, but he signed a ball for him anyway. Pomerance had Erdős sign the ball as well. By stretching the definition of joint publication a bit, Pomerance claims that Hank Aaron has an Erdős Number of 1.

The Erdős Number has itself become the focus of some semiserious mathematical study. Casper Goffman, a Purdue University mathematician, published a definition of the Erdős Number in 1969 that nicely shows how fuzzy concepts can be converted to unambiguous mathematics, and the price of the tradeoff:

Let A and B be mathematicians, and Let A_i, $i = 0,1,\ldots,n$, be mathematicians with A_0 = A, A_n = B, where A_i has written at least one joint paper with A_{i+1}, $i = 0,\ldots,n - 1$. Then $A_0, A_1, \ldots, A_n$ is called a chain of length n joining A to B. The A-number of B, $v(A; B)$, is the shortest length of all chains joining A to B. If there are no chains joining A to B, then $v(A; B) = +\infty$. Moreover, $v(A; A) = 0$. Then $v(A; B) = v(B; A)$ and $v(A; B) + v(B; C) \geq v(A; C)$.

For the special case A = Erdős, we obtain the function $v(\text{Erdős}; \cdot)$ whose domain is the set of all mathematicians.

As with all things mathematical, the definition of the Erdős Number has been refined over the years. As the number of mathematicians who could claim the honor of having an Erdős Number of 1 increased to a mob, a method of distinction among this already elite class had to be invented. "And now there is a new definition," Erdős liked to say. "If I have k joint papers with somebody his Erdős Number is one over k." The smaller the Erdős Number the closer to the master. The smallest Erdős Number is 1/57 and belongs to András Sárközy, who just edges out András Hajnal's 1/55.

Jerrold Grossman, a mathematician at Oakland University in Michigan, is the self-appointed guardian of the official Erdős Number list. Before Grossman came along, everybody talked about Erdős Numbers, but facts were hard to come by. While on a sabbatical Grossman "as a lark" began to compile the definitive list of Erdős collaborators and the people with whom they collaborated. Using a wide variety of sources, including voluminous bibliographies and "various necrological articles too numerous to list, and personal communications," Grossman pieced together the patterns of Erdős's vast and ever increasing network of collaboration. The arduous work had to be done by hand, since no computer could be trusted to distinguish between two mathematicians with the same name. "It was fun," Grossman says, "so I continued it."

Grossman publishes updates of his Erdős Number list every year and makes it available on his "Erdős Number Project" website, http://www.acs.oakland.edu/~grossman/erdoshp.html. He has continued to update his lists since Erdős's death. Because the prestige of an Erdős Number of 1 is so great, mathematicians have been dusting off and publishing old theorems proved with Erdős. As of February 1998, 485 people could officially claim the Erdős Number of 1; of those 193 have written more than one paper with Erdős. An additional total of 5,337 have the Erdős Number 2. Not even the energetic Grossman has tackled the enormous task of compiling a list of the legions who hold Erdős Number 3. According to Grossman's count, Erdős wrote 1,446 books, papers, and articles, a number he expects to top 1,500 by the time the time his remaining works are published and old ones are rediscovered.

Grossman's whimsical project has become a valuable source of insight into the sociology of mathematical collaboration. When Erdős started publishing, only a small percentage of all mathematics papers had more than one author. According to Grossman's figures, in 1940 about 90 percent of all mathematical papers were solo efforts; today that number has dropped to around 50 percent. Fifty years ago papers written with more than two people were almost unheard of, while today such multiple collaborations account for almost 10 percent of all published articles.

The majority of Erdős's early works, too, were solo efforts, but that quickly began to change. Erdős enlisted new collaborators at a pace that far outstripped the general trend. In total, only about one-third of his nearly 1,500 articles were written alone; 8 percent list four or more authors. Every year of his life Erdős added new collaborators, a trend that peaked in 1987, when he produced papers with thirty-five mathematicians with whom he had never before published.

Grossman is unable to account for the increasing tendency toward collaboration that has overtaken mathematics in the latter half of this century. Better communications probably is a large part of the reason. So in part is the growth of the publish-or-perish mentality that Erdős once satirized in verse:

> A theorem a day
> Means promotion and pay!
> A theorem a year
> And you're out on your ear!

The example of Erdős might also have helped drive the trend. An examination of Grossman's data shows that those who most frequently wrote papers with Erdős collaborated frequently with others. More than mathematics, Erdős's chief disciples learned from him a social style of *doing* mathematics that has become a norm.

Erdős bucked the stylistic trends of contemporary mathematics in other ways as well. Twentieth-century mathematics, Ernst Straus once pointed out, has been diomanted by "theory constructors," people who erect vast and general systems that illuminate mathematical structures. Erdős had an entirely different approach, focusing on specific problems, confident that as he solved them the general theory would slowly be revealed. He was, in the words of one friend, a "Socratic gadfly" who revealed truth by asking a series of careful questions. Erdős believed, in other words, that by the examination of a few carefully chosen trees a forest would be revealed.

"A theory is something that you can teach a whole course on," Joel Spencer once explained. "Individual questions are lower on the hierarchy, but I think Paul was the great exception to that." Erdős had an uncanny knack for picking the problems that reveal the overarching structure. A friend of Straus once complained to him in frustration that "Erdős only gives corollaries of the great metatheorems, which remain unformulated in the back of his mind." It was as if on some level Erdős already knew the theory but was capable of expressing himself only through specific questions. In a way, Erdős would always remain the brilliant child, the avid contributor of problems and solutions to *KöMal*, the monthly high school journal through which he had gained his first taste of the mathematical life and met his first mathematical friends.

Not everyone agreed with Erdős's approach to mathematics. Saunders MacLane spoke for those who disdained what he called the "Hungarian view of mathematics—that the science consists not in good answers but in hard questions." MacLane argued that Erdős's emphasis on problems was causing some mathematicians "to lose track of the fact that what matters most about a problem is its relevance." What MacLane failed to acknowledge was that, unlike many others, Erdős usually chose relevant questions, although their relevance to existing mathematical thought might take years to become obvious.

Erdős also had the ability to sense which problems could be solved and who might be able to solve them. Almost any mathematician can invent problems that are intractably difficult, or trivially simple, or whose solution leads nowhere. Erdős was a master at finding questions that lay on the shadowy border between those two domains, questions that were just hard enough and whose solutions suggested still other questions, opened doors, and spawned theories. What happens when you randomly add edges to a graph? It was a simple question that nobody had ever before asked; in the hands of Erdős and Rényi that question gave birth to an entire field of mathematics with far-reaching consequences.

According to his one-time assistant Straus, Einstein said that he did not become a mathematician because the field was so full of beautiful and difficult problems "that one might waste one's powers in pursuing them without finding the central questions." Erdős gave himself entirely over to the seduction that Einstein feared, but somehow he rarely spun off into irrelevancy. "This just proves to me," wrote Straus, "that in the search for truth there is room for Don Juans like Erdős and Sir Galahads like Einstein."

I N 1967 Joel Spencer ended his graduate studies at Harvard to work first at Bell Laboratories and then at the Rand Corporation in

Santa Monica, California. While at Rand he became fascinated by an Erdős problem concerning the best way to rank a tournament. In the tournament Erdős imagined, each participant plays a match with every other, and there are no ties. In a fair ranking, if player A is ranked above player B, then player A will actually beat player B. Unfortunately, no ranking system can be entirely fair; inevitably there will be upsets in which a lower-ranked player beats one with a higher rank. A fair ranking system is one that keeps the number of upsets to the minimum. Erdős wanted to know how fair the fairest ranking system could be.

To Spencer, Erdős was a legendary figure. "When you hear that something is an Erdős problem it just automatically puts a stamp on it," Spencer recalls. "That stamp was well earned; his problems were not selected at random. He had a style of always asking problems for which you were never really told explicitly the motivation. But these were never random problems, they were always the problems that really came to the frontier of the subject. To get the problem you'd have to do what Paul always called 'get a new idea,' you'd have to do something a little bit more. For one thing, you knew he hadn't solved them himself, so already you knew that puts it at quite a high level. My feeling even before I met him was that an Erdős problem was an enormous thing."

Spencer managed to come up with the crucial "new idea" necessary to solve Erdős's tournament problem. In those days Erdős frequently visited his friend Ernst Straus at UCLA, and on one of those visits Spencer arranged to meet with Erdős in his hotel room. "He was very welcoming," Spencer recalls. "If you had interest in mathematics he had this ability that some people have that it wouldn't matter if you were a high school student or my level or a senior person; if you had something interesting to say, then he would give you all of his attention." Erdős listened carefully as Spencer laid out his solution. When Spencer was through, Erdős nodded briefly and immediately launched into another problem. "This was the core of what he did," Spencer explained. "When he would talk to somebody he wouldn't try to push the person over to his area, he would look for the area that the person was already interested in and try to find a common bond." The solution to the Erdős problem that Spencer had brought with him to the hotel room became the core of his doctoral dissertation, and the new problem Erdős proposed led to the first of many joint papers. "In my case, and in the case of many of the people that I know, he could take people that already had worked in mathematics, already had shown some ability in mathematics, and just take them to an entirely new level."

Spencer came away from that first visit with more than a new problem to work on. He felt reassured that he had a place in the mathematical community, and he had a new sense of his mathematical vocation. Spencer recalls the feeling of being an awed newcomer to the world of mathematics. "When you're starting out, you see this world, and you want to be a part of it and there are these people way up there," seemingly inaccessible. "Not only would he talk with me, but we were doing mathematics! It was tremendous, we were proving and conjecturing, and it was the same when I was twenty-three as the year before he died. And it was the same with other people; if the mathematics was interesting, then he'd sit down and talk mathematics with you. That's a fantastic quality in terms of inspiring young people."

Erdős also inspired Spencer and many others with his total devotion to the subject. "Here was this person who was so brilliant and totally dedicated to doing mathematics, to its beauty," Spencer recalls thinking. "And here we were starting out, and we had seen glimpses of it, and we could see this senior person at the top of the field who was dedicating all of his energy toward it. To see him with his dedication, and to talk to him—you felt: This is great! This is absolutely what I want to be! I want to be part of this search! I mean, we already did want to be part, but he just moved us to a new intensity."

To Spencer and many other mathematicians, Erdős was a modern version of a medieval mendicant monk. Erdős is frequently called, without a trace of irony, a saint. Indeed, there was something saintly in Erdős's generosity, in his honesty and his support of the rights of the individual. But the essence of the saintliness his friends speak of was in his total devotion to the mathematical pursuit of pure beauty. Erdős often said that "property is a nuisance." In fact, to Erdős all aspects of life—jobs, money, property, and intimate personal attachments—that interfered with his devotion to mathematics were a nuisance to be avoided. While few people would choose to emulate him, Erdős's life was an example cherished by many.

The only exception to Erdős's pure focus on mathematics was his devotion to his beloved *Anyuka*. After the war, whenever he returned to Hungary, the two of them were almost inseparable. She was almost always present when Erdős took a new epsilon to lunch, and she accompanied him to the Mátraháza resort and elsewhere. She was a small, frail, dignified woman. "As a young woman she was pretty radical," Béla Bollobás, who met her when he was an epsilon in the late 1950s, recalls. "But by then she was rather the opposite. If someone had said she was from an aristocratic family you would have believed him." She was extremely proud of her famous son. "Mrs. Erdős lived for him," Bollobás recalls. "She collected his manuscripts and sent out his reprints." Erdős's cousin, Magda Fredro, once told an interviewer that Erdős's mother "saw in Paul the world. He was her God, her everything."

Erdős's devotion to his mother was no less complete. He worried about her health constantly. "He looked after her as somebody who needed looking after," Bollobás says. "When I knew her when she was in her eighties, she was a little frail, but certainly capable of taking care of herself, and Erdős was never a very worldly figure. In spite of that, they got on extremely well." It was a strange relationship, Bollobás allows, with its comic moments. After lunch in their apartment Erdős would invariably say, "*Anyuka*, don't you want to lie down?" Ten minutes later, after his mother had retired, Erdős would jump up, go to the bedroom and ask, "Aren't you asleep?" "This didn't happen just once, but every time," Bollobás said.

During her years of living alone in Budapest, *Anyuka* lamented that "children become letters." As she got older Erdős's mother increasingly missed her son when he was gone, so in 1964, at the age of eighty-four, *Anyuka* joined him on his perpetual journey. The only place to which she refused to accompany her son was India, because of her fear of disease. She made it clear to all that she had not chosen to accompany her son because she wanted to see the world. "I do not travel because I like it but because I want to be with my son."

Anyuka got her wish. She and Erdős ate every meal together, and she often sat in on his mathematical get-togethers, listening serenely. She worked on improving her English, since Erdős spent so much time in English-speaking countries, but never became fluent. Even if she could not understand everything that was said around her, she enjoyed the obvious respect with which Erdős was received wherever he went. "Being next to Erdős," Bollobás said, "she was very much the Queen Mother."

"They just understood one another completely," Anne Davenport, the wife of his old friend from Cambridge, recalled, "and he was absolutely devoted to her. It was most touching to see, the way he looked after her." A magazine story appeared about Erdős that claimed he held his mother's hand every night as she fell asleep. Erdős was offended when the story appeared and complained to a friend. "Well, don't you hold her hand?" the friend asked. "Yes," he admitted, "but not *every* night."

On a visit to southern California, Andrew Vazsonyi remembers renting a suite for Erdős and his mother in Westwood. "There was a nice living room and a bedroom," Vazsonyi recalls. "The manager explained that the lady would probably sleep in the bedroom and Erdős could open the sofa." Erdős's mother became oddly upset and began to complain about dust and threaten to leave. Erdős asked the manager if it would be possible to put a cot for him in the bedroom. "The problem disappeared," Vazsonyi said. "So I presumed he slept in the same room with her, and she would not tolerate his sleeping in the other room."

In 1971 Erdős and his mother visited Calgary, a frequent stop on his journeys. *Anyuka*, who had always enjoyed vigorous health, became suddenly ill and was admitted to the hospital. Hajnal, who was also in Calgary at the time, recalls that "Paul was actually quite impatient when she got ill." Erdős refused to believe that his ninety-year-old mother could be seriously sick and went to Edmonton to give a talk. "He didn't believe at the beginning that she was dying,"

Hajnal recalls. "Maybe the communication was bad with the doctors." Or maybe Erdős did not want to hear.

When Erdős returned from Edmonton his mother was no better. Hajnal tells of long days spent waiting with Erdős in the hospital corridors. "Paul wanted to talk mathematics all the time, to distract him from the actual happening. At one point, on the last day, I said no, no, let's not."

Erdős would never fully recover from his mother's death. He would always believe that the doctors had misdiagnosed her and that she should have lived longer. Erdős became lonelier and more depressed. "Strange," he would say, "I was always concerned when flying on a plane. But after my mother died I lost my fear." Erdős began to refer to himself as the PGOM, the Poor Great Old Man. Five years after her death a mathematician and friend named Herbert Wilf remembers meeting Erdős one morning and saying, "Hello, Paul, how are you today?"

"Herbert, I am feeling depressed this morning," Erdős replied.

"I'm sorry, Paul. Why is that?" Wilf asked, taken aback by Erdős's serious response to what was intended to be a meaningless social question.

"I miss my mother. She's dead, you know."

"I know that," Wilf said. "But that was five years ago."

"Yes, but I miss her very much."

After *Anyuka*'s death Turán counseled Erdős, "A strong fortress is our mathematics." Erdős began to work up to nineteen hours a day on mathematics. Strong coffee was inadequate fuel to run such a prodigious theorem-proving machine. For many years Erdős had occasionally taken amphetamines to help him work. Hajnal recalls that as early as 1957 Erdős had had access to Benzedrine. "I took some of those pills," Hajnal says. "When I took it I could work eighteen hours in a row." But unlike Erdős, Hajnal did not have a steady supply.

"He was very disciplined with it," Hajnal said, and took it only in relatively small doses on days he wanted to work particularly hard. With his mother gone, Erdős wanted to work hard all the time and began to take amphetamines, washed down with strong coffee, every day. Now, whenever he stopped to admire an epsilon, he had a new trick: He would reach into his coat pocket and pull out a bottle of amphetamines, which he held out in front of him clenched in his fist. "Watch this," he would say, and suddenly spread his fingers wide. The bottle plummeted toward the ground, to be snatched up at the last possible instant, demonstrating reflexes that became ever more impressive as Erdős aged.

When she was alive, Erdős's mother had kept all of his thousands of reprints and his voluminous correspondence neatly filed in their Budapest apartment, sending out copies to whoever requested them. After her death Erdős refused ever again to stay in the apartment that he and his mother had jointly owned; when he was in Budapest he would stay in an apartment at the Academy of Sciences. In the years of his travels with *Anyuka* the chore of maintaining his papers and keeping track of his finances and the other details of his peripatetic life had fallen mostly on the athletic shoulders of an American mathematician named Ronald Graham.

Ron Graham and Paul Erdős were the odd couple of mathematics, so different from each other that their friendship was perhaps inevitable. Graham was born in 1936 in Taft, California, a small town northwest of Los Angeles. He spent most of his youth crisscrossing the United States as his father moved between oil fields and shipyards in California and Georgia. "Because I was always kind of a new kid," he recalls, "I was never accepted socially into the in-group." He skipped a few grades, which didn't help either, because he was always younger and smaller than his classmates.

So Graham concentrated on his two main interests, mathematics and astronomy. The stars still interest him, but, thanks to some good teachers and the fact that "math is portable," he chose mathematics. When he was fifteen he won a scholarship to attend the University of Chicago.

In those days the University of Chicago was experimenting with the Chicago Plan, a liberal education devised by Robert Maynard Hutchins and based upon reading the Great Books. When he entered the university, Graham took a series of tests, which indicated that while he was very advanced in mathematics and science, he was deficient in literature, sociology, philosophy, and the other liberal arts.

Graham was required to make up for those deficiencies and was not allowed to take any mathematics. So Graham read the Great Books and in his spare time began to learn gymnastics, a sport where his still small size—he would later grow to a strapping 6 feet 2 inches—was not a disadvantage. Graham never does anything by halves: He became a professional trampolinist and master juggler able to solve three Rubik's cubes while juggling them. But after three years without mathematics, Graham began to suffer from withdrawal and left to attend the University of California at Berkeley.

At Berkeley Graham enrolled as an electrical engineering student and found time to complete a thesis with the great number theorist D. H. Lehmer. Unfortunately, by that time Graham had been a student for so long that his deferment had expired and he was eligible for the draft. To avoid being drafted he enlisted in the Air Force for a four-year tour and was promptly sent to Alaska. "I was able to work nights and go to school in the daytime," Graham says. The only problem was that the University of Alaska was not accredited in mathematics, so Graham settled for a physics degree, or almost. Before graduation, in 1958, the Air Force shipped Graham overseas.

When Graham's tour of duty was over, he returned to Berkeley finally to concentrate full-time on mathematics. While working on his doctorate Graham showed his adviser a little problem he had been fooling around with. His adviser suggested to Graham that he write to Erdős, who was interested in that kind of thing. Graham had never really heard about Erdős—"I was out of touch; I'd been in Alaska" —but he wrote anyway. Erdős wrote back quickly, a typical note that probably began: "That's an interesting question. Consider the following generalization . . ."

Graham would not meet Erdős in person for another couple of years. After receiving his PhD from Berkeley, Graham went to work for Bell Labs. In contrast to Erdős, who never had a job that lasted more than an academic year, Graham would stay at Bell for his entire career, eventually becoming Chief Scientist of AT&T Labs, as it is now known. Shortly after beginning at Bell Labs, in 1963, Graham attended a number theory meeting in Boulder, Colorado, which was also the first meeting Erdős attended in the United States since Sam had decided he was not a security threat. "Ron had about twenty papers he had written before he had taken the job at Bell Labs, and they were all coming out," John Selfridge, another friend and collaborator, recalls. "We sat there listening to all his stuff, and it was just all so good."

What Graham recalls most vividly about his first meeting with Erdős is not some elegant proof or provocative conjecture, but a game of ping-pong. Erdős loved the game and had spent countless hours playing it with Turán and his friends at Mátraháza. Small, frail, bespectacled, absent-minded, and given to bouts of dizziness, Erdős with a paddle in his hands was not an imposing figure. "As soon as Erdős took his stand behind the ping-pong table, it was clear that he was an amateur and could not be a serious opponent," Janos Pach recalled. He held the paddle gingerly, as if afraid of catching a disease from it, and his serves were "totally ridiculous; it was hard not to smash them." Erdős challenged Graham—6 feet, 2 inches, and athletic, with the reflexes of a trained acrobat—to a game of ping-pong. Graham lost.

Erdős's victory was not a fluke. Despite his obvious shortcomings Erdős had remarkably fast reflexes. "One could not help thinking that in his nervous system impulses somehow traveled a lot faster," Pach said. Without undue motion Erdős could return most shots, neutralizing his opponents. "He played very defensively; he wasn't bad," Graham says. "That's how I got into ping-pong. Because he could beat me. He beat me and I didn't see how that was possible. He was an old guy."

Graham does not like to lose, especially when it can be avoided. When he got back to Bell Labs he hired a ping-pong coach, bought a machine that fired endless rapid volleys of spinning ping-pong balls at him, and practiced. Before long he was Bell Labs' ping-pong champion. Employing the same systematic, single-minded approach, Graham has mastered bowling (he's rolled several perfect games), Chinese (on the phone he passes for a native), the piano, and countless tricks with balls, coins, and cards—all while keeping a schedule that seems designed to kill: Besides his duties at Bell Labs and churning out mathematical books and papers (he's written more than two hundred papers plus several books), he is on the editorial boards of more than forty mathematical journals, lectures constantly, sits on prestigious government committees, is a treasurer of the National Academy of Sciences and the National Research Council, coaches gymnastics, and somehow never seems to be rushed. Asked how he manages to accomplish all this, Graham replies in his slow, quiet voice, "Well, there are a hundred and sixty-eight hours in a week."

For thirty years and more, a few of those hours were spent handling the financial and administrative details of Erdős's rootless life. "He never had a checking account," Graham explains. "I personally liked to carry, not a lot of money, but five hundred or a thousand dollars just because you're traveling and it's a good habit. And [Erdős] discovered that soon enough."

"Ron, lend me some money," Erdős would say.

"How much do you need?"

"Well, how much can you afford?"

So Graham would lend Erdős $300. Erdős would always repay what he borrowed, but without a bank account his cash flow could be uneven. After receiving a loan from Graham, Erdős would sometimes ask for a hundred dollars more for a Hungarian friend. "At that time some of the Hungarians were planning to leave the country and were trying to establish hard currency accounts in the West. It wasn't quite legal then," Graham explains. Erdős managed to get Graham to help out.

Eventually Graham got Erdős a checking account, into which he deposited all the checks Erdős received for his various speaking engagements and honoraria from all around the world. "It seemed an easy thing to do. After a while it became automatic," Graham says. Since Erdős was almost impossible to get hold of most of the time, Graham learned how to forge Erdős's signature in order to endorse his checks. "I had copies of his handwriting and I would practice," he says. "I thought to myself: Think shaky. And I got it down pretty well. But over the years my version of his signature and his version tended to diverge pretty much. I'd be willing to bet that the bank wouldn't accept a check that he endorsed."

For many years Erdős had made a practice of offering cash prizes for the solutions to problems that had begun to bother him. He didn't care who solved them, he just wanted them solved and decided that a cash bounty was the best way to achieve the result. "Like the kings of old," Straus wrote, Erdős "offer[ed] differing cash rewards for the different levels of difficulty, so that we [got] not only the question but also his assessment of the task it poses." Some of his simpler problems went for as little as a dollar or two, but he offered as much as $10,000 for problems he believed to be "hopeless." The largest prize Erdős ever had to pay off was $1,000. "Someone once asked me what would happen if all the problems were solved at once," Erdős once said. "Could I pay? Of course I couldn't. But what would happen to the strongest bank if all the creditors asked for their money back? The bank would surely go broke. A run on the bank is much more likely than solutions to all my problems."

Still, Erdős frequently had to pay off his prizes. For years Graham kept books of checks drawn on long-closed accounts that Erdős would use to pay off his prizes. It did not matter that the checks were not genuine, since a framed check from Erdős was worth far more to mathematicians than the check's nominal value. But after a while it began to occur to people that a framed *canceled* check, which they could obtain from Graham, would also look good, and they could pocket the cash. So Graham had to start using checks on valid accounts.

In his life Erdős paid off three or four thousand dollars' worth of prizes, but many more of his problems remain unsolved. Graham and a few of Erdős's other friends have vowed to pay any outstanding prizes for solutions. "Many of these problems have been around for a long time and probably should go up in value," Graham says, though, given the size of the fiscal responsibility he has already assumed, it's unlikely that he will increase them.

Erdős became a frequent visitor to Bell Labs in Murray Hill, New Jersey, spending as much as a month every year at the home of Graham and his wife, the mathematician and Erdős collaborator Fan Chung. In the late 1980s Graham and Chung built a special "Erdős Room" with its own private bath and telephone, which Erdős especially liked because, thanks to Graham's corporate perks, he could freely call anywhere in the world. Erdős also had access to a library with the latest journals and the filing cabinets in which Graham stored reprints of all his papers and much of his correspondence. It was as much an act of survival as of generosity. Erdős was a very difficult house guest.

There is some debate about whether Erdős was truly incapable of mastering even the simplest domestic chore, or whether his incompetence was feigned. In small doses Erdős's helplessness could be endearing, but the charm of a visit soon wore thin. Everyone with whom Erdős stayed has tales to tell of containers of tomato juice dripping on refrigerator shelves and apparently unopenable bags of pretzels. "He does not like to be alone very much," Fan Chung said. "When he stays at our house not only do we have one guest but in fact we have all the time a stream of so-called Paul-sitters!" Erdős could not drive, so his hosts became chauffeurs; he could not sew, so his hosts became tailors; he could not pack a suitcase, so his hosts became valets. Mostly his hosts just became tired. "I was very happy to see him come when he came," Carl Pomerance said, "and happy to see him go when he went." The only people who would not agree were the epsilons, the children of Erdős's hosts, who always looked forward to a visit from Uncle Paul.

Erdős usually stayed only about a week or so at any one place before he had thoroughly exhausted the brains and patience of all local mathematicians and had to move on in search of further stimulation. Once, while visting Stanford University, he had been camped in the home of his friend Gabor Szego's house and gave no sign of imminent departure. Szego's wife met Vazsonyi at a party one night and said in despair, "Erdős dropped in three weeks ago and he is still staying with us. I am at the end of my wits." Vazsonyi told her, "No problem. Tell him to get out."

"I couldn't do that," she said. "We love him and could not insult him."

"Do what I said," Vazsonyi insisted. "He will not be insulted at all."

An hour later Erdős came up to Vazsonyi and asked for a ride to a hotel. "What happened?" Vazsonyi asked innocently. "Oh, Mrs. Szego asked me to move out because I stayed long enough," he said, totally undisturbed.

Another aspect of Paul-sitting involved looking after Erdős's health. Graham, like many of Erdős's friends, was concerned with Erdős's daily consumption of amphetamines. In 1979, in an attempt to wean Erdős from the drugs or at least to prove to him that he had a problem, Graham bet Erdős $500 that he could not refrain from taking speed for a month. Erdős won the bet and then promptly resumed popping pills. "You showed me I'm not an addict," Erdős said. "But I didn't get any work done. I'd get up in the morning and stare at a blank piece of paper. I'd have no ideas, just like an ordinary person. You set mathematics back a month!"

Aside from that one month, Erdős maintained his extraordinary pace, filling blank pages with theorems and conjectures, well past the age when most mathematicians write their memoirs. For many years the only signs of his aging were the slow accretion of letters that he would add to his name, PGOM, with two new letters added every five years until he became, at seventy-five, PGOMLDADLDCD.* He recited those initials at lectures—which he inevitably called "sermons"—all over the world, standing on the podium with his arms folded across his stomach, his hands grasping his elbows, speaking slowly and quietly with a canny comic timing. "He was the Bob Hope of mathematics," his friend Melvin Nathanson liked to say. Erdős would tell his stories about the crazy Sidon, and the brilliant young Pósa, and how he himself had discovered negative numbers at the age of three. And he made jokes about aging: Next year you'll hold this conference in my memory; it seems that I was once young, but that was long ago. He told about calling to congratulate his friend Polya on his ninety-seventh birthday. "I told him you will celebrate your hundredth birthday with great splendor. He said, maybe I want to be a hundred, but not a hundred one because old age and stupidity are very unpleasant." Then he would say, "Let me stop talking nonsense about the past and start talking about mathematics" and get down to work describing his favorite problems and how much he was willing to pay for their solutions.

* Poor Great Old Man, Living Dead, Archaeological Discovery, Legally Dead, Counts Dead.

By the time he had added the final two letters to his signature at the age of seventy-five, Erdős's finest work was behind him. His ideas would no longer result in the explosion of new branches of mathematics, such as Ramsey theory, combinatorial geometry, extremal graph theory, or the theory of random graphs, or in the creation of powerful mathematical tools like the probabilistic method, but the pace of his publications did not appreciably slow. Every year he circled the globe, picking up new collaborators while periodically revisiting his friends, like a migratory bird. One of his favorite stops each year was Western Michigan University in Kalamazoo, one of the world's leading centers of graph theory. When he was there he would stay with Yousef Alavi, a mathematician who expressed his exasperation by shouting, "This is highly irregular!" Alavi had always wanted to write a paper with Erdős but had been unable to come up with a suitable topic. One day Alavi used his expression in the presence of Erdős, Graham, and some others. Graph theorists often analyze objects called regular graphs, which are graphs in which the same number of edges are incident on each vertex (a square is a regular graph, with two edges incident on each vertex). Prompted by Alavi's expression, the mathematicians wondered what a "highly irregular" graph would be. After some discussion they came up with a suitable definition and began to prove theorems. The whimsically inspired definition proved to be interesting and the result was a series of large joint papers defining an interesting new area of study. And Alavi achieved his dream of acquiring an Erdős Number of 1.

In the late 1980s Erdős developed a heart condition. Alavi introduced him to a cardiologist named Janos Gellert. "He was an absolutely wonderful man," Gellert recalls of Erdős. "He was interested in everything. You felt right away that you are not dealing with one of your colleagues or an average guy. He was a genius, his thoughts were all over the place. I've met very smart people. I have never met a genius before."

Genius or not, Erdős was a terrible patient. He would amaze Gellert with his knowledge of the latest medical techniques and theories and could talk medicine for hours, but, as in most other aspects of his life, he did not particularly like being told what to do. Gellert tried to discourage Erdős from taking amphetamines, but to no avail. He prescribed anti-arrhythmia medication to be taken every eight hours; Erdős popped a few whenever he felt like it. "He basically disregarded any disciplined approach to anything," Gellert said. Erdős tried to reassure Gellert by telling him that he always took stairs and never used elevators. In the middle of dinner at Gellert's house Erdős would spring up out of his chair and run up and down the stairs to prove his fitness. "I just wanted to show the doctor the good shape I'm in," he would say. But as the evening wore on Erdős would frequently nod off in his chair while the conversation flowed around him. Sometimes he fell off the sofa or the chair. Once he asked Gellert's wife for some espresso, then nodded off. When a guest said, "Don't bother, he's asleep," Erdős suddenly sat upright and shouted, "I'm not sleeping!"

Erdős worried about his health but hated to give up any of the time he devoted to mathematics, even for serious health problems. Erdős was a frequent visitor to the University of Memphis, where he would work with the mathematician Richard Schelp and others and get treatment for his serious eye problems. Schelp remembers working with Erdős in his hospital room once when a nurse came in. "What are you doing?" she asked. Erdős took a few minutes to explain the basics of prime numbers to the nurse. When she returned to the room later, Erdős quizzed her on their previous conversation. After she left, he said to Schelp, "Either she's not so clever or I'm not a good teacher."

Erdős needed a cornea transplant to restore vision to one of his eyes. As he was leaving Memphis, a donor became available. At first Erdős did not want to delay his travels for the operation, but after a lengthy argument his friends convinced him that his eyesight was more important. Nevertheless, he insisted on taking a pad with him to the operating room to continue his calculations. When the surgeon saw this he said, "You won't need that. I'll be working on your eye." Erdős replied, "I'll do math with the other eye."

Erdős complained to Dr. Gellert, as he did to everyone, that he was depressed all the time. Even more troubling to him was that he was not as alert as he had been. Colleagues had begun to notice. Pomerance mentioned a problem to Erdős, "and he said, 'oh, that's a very nice problem.' A month later on the phone he mentioned the same problem to me. This wouldn't have happened when he was in his prime." Erdős wrote a number theory paper with colleagues in Israel in which he improved on a previous result that he and Bollobás had discovered. Unfortunately, Erdős had entirely forgotten this previous work and included the result in the paper as if it were new. When Bollobás pointed this out, Erdős was extremely upset. "Who cares?" Bollobás said. "I care, of course," Erdős insisted. "I really care." Compared with most mathematicians, Erdős was still extraordinarily fast, but nothing like what he was in his prime. A young mathematician named Neil Calkin, one of Erdős's last collaborators, says, "One of my greatest regrets is that I didn't know him when he was a million times faster than most people. When I knew him he was only hundreds of times faster."

Erdős still traveled at what for others would seem a furious pace, but the invitations were beginning to dry up. His visits with his friends were getting longer and more difficult. He called the Szekereses in the middle of the Australian night, as usual disregarding time zones, and asked to come for an extended visit. Esther Szekeres was not in very good health, their home was small, and the prospect of Erdős's disruptive company, with his favorite noise playing at all hours (which was, oddly, not Bach but Ravel's *Bolero*) did not thrill George Szekeres. But after a discussion with Esther, he invited Erdős for a long visit.

Even at his reduced pace, during the last months of 1995 and early 1996 Erdős had made visits to Atlanta, Memphis, three cities in Texas, Bell Labs and Rutgers University in New Jersey, New Haven, Baton Rouge, Colorado, France, and Germany. In February 1996 Erdős was in Kalamazoo to attend an international graph theory conference. Toward the conclusion of one of the principal talks, Allen Schwenk of Western Michigan University noticed that Erdős, who was sitting in the second row, did not look well. Others had already noticed and were supporting Erdős when suddenly his head dropped to his chest and he went limp. An ambulance was called, and Erdős was rushed to the emergency room.

In the emergency room Erdős regained consciousness and was

unworried. He answered the doctors' questions impatiently and turned his attention to the mathematicians who had accompanied him, Schwenk, Ron Graham, Ralph Faudree, and John Selfridge. "Ralph," he said, "I was thinking about that problem we discussed earlier. Have you tried this approach?"

Alavi arrived and then Dr. Gellert, who quickly examined Erdős. By now Erdős was walking around the room, and when Gellert asked if Erdős minded discussing his condition with so many visitors, Erdős waved his hands and said, "Of course it's okay. These are my friends."

According to Dr. Gellert, Erdős was suffering from "sick sinus syndrome," a condition that caused his heart rate to drop dangerously. Gellert told him that he needed to have a pacemaker inserted right away, which would require a minor surgical procedure and at least a day in the hospital. "He looked at me like I was out of my mind," Gellert recalled. He looked at his watch and said, "I have to be at the conference. Plus this is the evening of the banquet! Impossible, absolutely impossible!" Erdős explained that he was scheduled to go to Philadelphia the following week, and then to Israel. "Maybe when I return to Hungary I can have it done."

"You take a great chance," Gellert warned. "Not only might you pass out, but you might not come to again." Erdős agreed to have a pacemaker installed on the condition that it be done right away and he could still attend the banquet. "Well, normally you have to fast, you have to take sedatives and you put a fresh lead into the heart, so we keep people in the hospital for the first twenty-four hours," Gellert explained. But it was clear that the only way Erdős would consent would be if it were done right away. In Gellert's mind, the risks of not implanting the pacemaker outweighed the risks of rushing it. He agreed to Erdős's terms but insisted on attending the banquet himself, accompanied by another cardiologist. Gellert called a colleague and scheduled the procedure.

When Erdős was brought into the catheterization lab for the insertion, he refused sedatives. "It was like restraining a wild animal. It was not like him to be limited in any way with his activities or anything," Gellert said. Finally, after the pacemaker was in, Gellert got

Erdős to bed and with Alavi's help, talked him into staying for a few hours.

At the banquet Erdős got up and said, "I always make this joke. 'You can hold the next conference in my memory.' This time you almost did." He then launched into a sensible discussion of a point made by the speaker at the moment he had passed out. Within minutes, despite Gellert's warning to keep his arm, which was in a sling, motionless, Erdős was gesturing emphatically. The next day he walked for more than an hour with Alavi in a bird sanctuary.

Erdős kept his commitments and raced on to Philadelphia, Israel, and Hungary. Melvin Henriksen recalls that Erdős attended a talk he gave at the Hungarian Academy of Sciences in early September, interrupting with his usual sharp comments. When Henriksen said goodbye Erdős was working with a young mathematician on a paper. Erdős had frequently repeated his parody of a well-known couplet of Hungarian poetry:

> One thought disturbs me, that I may decease
> In slowly progressing Alzheimer's disease.

He clearly escaped this tragedy. The Erdős Henriksen had taken leave of in Budapest was much like the man he had met many years earlier at Purdue. At that time Erdős lamented to all he met, "Death begins at forty." But forty years later Erdős's brain was still open.

From Budapest Erdős flew to Warsaw for a combinatorics meeting, where he gave two talks. Early on Friday morning, September 20, Erdős suffered a heart attack alone in his room. He was carried to a hospital, where he suffered a second, fatal heart attack later that afternoon. Erdős was eighty-three when he "left."

Erdős had joked so frequently over the years about his ever looming departure that as news of the event spread around the world via the Internet and phone calls, there was a momentary sense of unreality.

Erdős had very nearly achieved his often stated desire to "die with his boots on." He liked to tell a story about the death of another prolific mathematician: "Euler, when he died, simply collapsed and said, 'I am finished.' [One time] when I told this story somebody callously remarked: 'Another conjecture of Euler's proven.'"

Erdős had a conjecture about his own death: He would be giving a lecture announcing one more startling new result. A voice from the audience would shout, "Yes, but what of the general case?"

In his fantasy, Erdős would reply, "I leave that to the next generation," and then he would leave. Erdős would have been gratified that his actual death came very near to proving his conjecture.

T H E passing of most mathematicians, like their lives, usually goes unnoticed outside of their narrow community. But when Erdős died, the front page of *The New York Times* announced, "Paul Erdos, 83, a Wayfarer in Math's Vanguard, Is Dead." All around the world newspapers ran long obituaries telling of Erdős's brilliance and the importance of his work, emphasizing his eccentricities. The political columnist Charles Krauthammer read those obituaries and, like many friends of Erdős, was disturbed by how little noticed were Erdős's essential goodness and generosity. He was particularly offended by the abrupt conclusion of the *Washington Post* obituary, which stated painfully that Erdős "leaves no immediate survivors." Krauthammer countered with a story about Erdős—one of a hundred similar stories —that he had heard from Graham. Graham repeated his account in the speech he gave at Erdős's memorial service in Budapest.

A talented young mathematics student Graham knew, Glen Whitney, had been accepted at Harvard but could not afford to go. Whitney had scraped together what money he could but was still short. One day Graham mentioned Whitney's story to Erdős, who was visiting. Erdős arranged to see Whitney and loaned him $1,000, which was a lot for a man who rarely carried more than thirty dollars in his pocket and had never accumulated much in the way of savings. Erdős told Whitney that he could pay him back when he was able.

A few years later Whitney had graduated from Harvard and was teaching in Michigan. He contacted Graham with the news that he could now afford to repay Erdős. Graham told Erdős, who merely said, "Tell him to do with the thousand dollars what I did."

"No survivors, indeed," Krauthammer concluded.

● ● ●

O N October 18, a month after Erdős died, several hundred of his survivors gathered together on a cold autumn morning in Budapest's Kerepesi Cemetery to pay their final respects. The mathematicians were of all ages and had come from around the world, but white heads dominated. Many of Erdős's friends had not seen each other in years but had kept in touch through his tireless wanderings. They greeted one another sadly, suddenly aware that with Erdős's leaving, an age was passing. "Well, I won't say you look good anymore, André," said one elderly mathematician greeting a stooped friend, "but you're alive, which is something." Another lamented that "our weighted Erdős Numbers [which vary inversely with the number of papers written with Erdős] will no longer decrease." A relatively young mathematician sobbed when he said, "I was afraid the Supreme Fascist would take him sometime, but I was always hoping it would not be so soon."

In the modern chapel the mathematicians gathered around the urn containing Erdős's ashes and one by one made speeches of farewell. Graham quoted the maxim carved on David Hilbert's gravestone, which he felt also expressed the force that drove Erdős: *"Wir müssen wissen / Wir werden wissen."* We must know. We will know. As the wet leaves fell from the great trees in Kerepesi Cemetery, Erdős's first collaborator, George Szekeres, who had returned to Budapest only once before in fifty years, expressed what was in so many hearts. "To me," Szekeres said, "a farewell to Paul means a farewell to the whole world of our young years, to the mathematical discussions under the Anonymous memorial in the City Park, to the weekend excursions. . . . In the world of mathematics thousands will grieve in their own individual ways the gap that Paul's departure has created for all of us."

Never again the knock at the door or the midnight phone call. Never again the bold announcement, the challenge and promise: "My brain is open!"

I never had the pleasure of meeting Paul Erdős, so I had to rely on the memories of the many people who have. Fortunately, Erdős had hundreds of collaborators and friends and all of them have vivid memories. To write this book I primarily relied on interviews with many of the people who knew Erdős best.

My first source for information on all things relating to Paul Erdős was his closest American friend, Ronald Graham. After Erdős's death, when I began to consider writing this book, Graham suggested I attend Erdős's memorial service in Budapest. Erdős's closest Hungarian friend and collaborator, Vera Sós, was kind enough to invite me to attend the memorial and help me on my visit to Budapest.

The wonderful story-telling ability of Andrew Vazsonyi, one of Erdős's oldest friends, helped bring Erdős to life for me. I spent several days interviewing Vazsonyi and drew heavily from his written memoirs of his friend. I also visited George and Esther Szekeres in Sydney, Australia, and they generously shared memories of their lifelong friendship with Erdős.

To recapture Erdős's voice and for anecdotes, facts, and a detailed timeline of Erdős's life, I drew heavily on László Babai's wonderful biographical essay, "Paul Erdős is Eighty." Babai interviewed Erdős extensively for his essay, which Erdős personally checked for accuracy. Bela Bollobás, Melvin Henricksen, Janos Pach, Alan Schwenk, and Joel Spencer all wrote fascinating memoirs of Erdős. I made extensive use of these memoirs (and many shorter memoirs that appeared on the Internet) and of subsequent interviews with the authors. Magazine articles by John Tierney and Paul Hoffmann were also valuable sources of information.

George Csicsery's beautiful documentary, *N is a Number*, allowed me to see the living Erdős. Csicsery also released several of Erdős's lectures on videotape, from which I came to understand why Erdős is sometimes called "the Bob Hope of mathematics." In the year following Erdős's death there were many memorial sessions in which those who knew him best shared their memories of his life and work. I attended one of the largest of these, which was held during a meeting of the Mathematical Association of America in Atlanta, Georgia. The talks, panel discussions, and corridor conversations helped give me a sense of the impact of Erdős's spirit on the mathematical community. The details of Erdős's controversy with Alte Selberg over the elementary proof of the Prime Number Theorem were largely unknown until very recently. Dorian Goldfeld kindly sent me a preprint of a paper he wrote on this incident based on his unique access to letters and documents that have been unpublished for fifty years. Melvyn Nathanson shared his insights into this event along with much other useful and fascinating information.

The archives of the Institute for Advanced Study and University of Notre Dame provided documents that helped clarify some of Erdős's perpetual comings and goings. I also conducted dozens of other interviews to complete my picture of Erdős. The names of the major figures interviewed are listed in the Acknowledgments.

Aczel, Amir D. *Fermat's Last Theorem: Unlocking the Secret of an Ancient Mathematical Problem.* New York: Four Walls Eight Windows, 1996.

Albers, Donald J., and G. L. Alexanderson, eds. *Mathematical People: Profiles and Interviews.* Boston: Birkhäuser, 1985.

Babai, László, and Joel Spencer. "Paul Erdős (1913–1996)." *Notices of the American Mathematical Society,* vol. 45, no. 1, January 1998, pp. 64–73.

Babai, László, et al. "The Mathematics of Paul Erdős." *Notices of the American Mathematical Society,* vol. 45, no. 1, January 1998, pp. 19–32.

Baker, A., et al., eds. *A Tribute to Paul Erdős.* Cambridge: Cambridge University Press, 1990.

Barson, Michael. *Better Red Than Dead: A Nostalgic Look at the Golden Years of Russiaphobia, Red-baiting, and Other Commie Madness.* New York: Hyperion, 1992.

Beiler, Albert H. *Recreations in the Theory of Numbers: The Queen of Mathematics Entertains.* New York: Dover Publications, Inc., 1966.

Bell, E. T. *Men of Mathematics.* New York: Simon & Schuster, 1937.

Bellman, Richard. *Eye of the Hurricane: An Autobiography.* Singapore: World Scientific, 1984.

Bernstein, Peter L. *Against the Gods: The Remarkable Story of Risk.* New York: John Wiley & Sons, Inc., 1996.

Berzsenyi, George. "In Memoriam: Paul Erdős (1913–1996)." *Quantum*, November/December 1996, pp. 40–41.

Cole, K. C. *The Universe and the Teacup: The Mathematics of Truth and Beauty.* New York: Harcourt Brace & Company, 1998.

Conway, John H., and Richard K. Guy. *The Book of Numbers.* New York: Copernicus, 1996.

Csicsery, George Paul. "*N Is a Number: A Portrait of Paul Erdős.* A documentary film (57 min). Oakland, CA, 1993. Video available from the Mathematical Association of America.

Csicsery, George Paul. *To Prove and Conjecture: Excerpts from Three Lectures by Paul Erdős.* A documentary film (53 min). Oakland, CA, 1989. Video available from the Mathematical Association of America.

Dauben, Joseph Warren. *Georg Cantor: His Mathematics and Philosophy of the Infinite.* Princeton: Princeton University Press, 1979.

Davis, Philip J., *Mathematical Encounters of the 2nd Kind.* Boston: Birkhäuser, 1997.

Davis, Philip J., et al. *The Mathematical Experience.* Boston: Birkhäuser, 1981.

Dembart, Lee. "Mathematician Erdős: In the World of Numbers, He's No. 1." *Los Angeles Times*, March 26, 1983, p. 1.

Devlin, Keith. "Bring Home the Bacon." *The Guardian*, February 12, 1998, p. 8.

Dickson, Leonard Eugene. *History of the Theory of Numbers.* Volume 1: *Divisibility and Primality.* New York: Chelsea Publishing Company, 1952.

Donald, David Herbert. *Lincoln.* New York: Simon & Schuster, 1995.

Dörrie, Heinrich. *100 Great Problems of Elementary Mathematics:*

Their History and Solution. Trans. David Antin. New York: Dover Publications, Inc., 1958 (translated in 1965).

Dunham, William. *The Mathematical Universe: An Alphabetical Journey Through the Great Proofs, Problems, and Personalities.* New York: John Wiley & Sons, Inc., 1994, pp. 1–12.

Edsall, John T. Letter to House Judiciary Committee. Excerpted in *Science,* May 14, 1954, p. 677.

Erdős, Paul. *The Art of Counting: Selected Writings.* Ed. Joel Spencer. Cambridge, Mass.: The MIT Press, 1973.

Erdős, Paul. "Child Prodigies." *Proc. Washington State Univ. on Number Theory.* Ed. J. J. Jordan and W. A. Webb. Pullman, Wash.: Washington State University, 1971.

Erdős, Paul. "My Joint Work with Richard Rado." *Proc. 11th British Combinat. Conf.* Cambridge: Cambridge University Press, 1986, pp. 53–80.

Erdős, Paul, and Carl Pomerance. "On the Largest Prime Factors of n and $n + 1$." *Aequationes Mathematicae,* vol. 17 (1978), pp. 311–321.

Erdős, Paul, and A. Rényi. "On the Evolution of Random Graphs." *Magyar Tud. Akad. Mat. Kutató Int. Közl,* vol. 5 (1960), pp. 17–61.

Erdős, Paul, and Joel Spencer. *Probabilistic Methods in Combinatorics.* Budapest: Akadémiai Kiadó, 1974.

Erdős, Paul, and Georg Szekeres. "A Combinatorial Problem in Geometry." *Compositio Math.,* vol. 2 (1935), pp. 463–470.

Gardner, Martin. *Fractal Music, Hypercards, & More.* San Francisco: W H Freeman & Co., 1991.

Gardner, Martin. *Penrose Tiles to Trapdoor Ciphers: And the Return of Dr. Matrix.* Washington, D.C.: MAA Spectrum, 1997.

Gardner, Martin. *Wheels & Other Mathematical Amusements.* San Francisco: W H Freeman & Co., 1985.

Goffman, Casper. "And What Is Your Erdős Number?" *American Mathematical Monthly,* vol. 76 (1969), p. 791.

Goldfeld, D. "The Elementary Proof of the Prime Number Theorem: An Historical Perspective." Unpublished paper.

Graham, R. L., and J. Nešetril, eds. *The Mathematics of Paul Erdős.* New York: Springer-Verlag, 1997.

Hadamard, Jacques. *The Psychology of Invention in the Mathematical Field.* New York: Dover Publications, Inc., 1945.

Halmos, Paul R. *I Want to Be a Mathematician: An Autobiography in Three Parts.* Washington, D.C.: Mathematical Association of America, 1985.

Harary, Frank. "In Praise of Paul Erdős on his Seventieth." *Journal of Graph Theory,* vol. 7 (1983), pp. 385–390.

Hardy, G. H. *A Mathematician's Apology.* Cambridge: Cambridge University Press, 1940.

Hoffman, Paul. "The Man Who Loves Only Numbers." *Atlantic Monthly,* vol. 260, November 1987, p. 60(7).

Honsberger, R. *From Erdős to Kiev: Problems of Olympiad Caliber.* Washington, D.C.: Mathematical Association of America, 1996.

Honsberger, R. *Mathematical Gems III.* Washington, D.C.: Mathematical Association of America, 1985.

Honsberger, R. *Mathematical Morsels.* Washington, D.C.: Mathematical Association of America, 1978.

Johnson, Paul. *Modern Times.* New York: Harper & Row, 1985.

Kac, Mark. *Enigmas of Chance: An Autobiography.* Berkeley: University of California Press, 1974.

Kac, Mark, and Stanislaw M. Ulam. *Mathematics and Logic.* New York: Dover Publications, Inc., 1968.

Kac, Mark, et al. *Discrete Thoughts: Essays on Mathematics, Science and Philosophy.* Boston: Birkhäuser, 1992.

Kanigel, Robert. *The Man Who Knew Infinity: A Life of the Genius Ramanujan.* New York: Washington Square Press, 1991.

Kauffman, Stuart. *At Home in the Universe: The Search for the Laws of Self-Organization and Complexity.* New York: Oxford University Press, 1995.

Klein, Felix. *Famous Problems of Elementary Geometry.* New York: Dover Publications, Inc., 1930.

Kolata, Gina Bari. "Mathematician Paul Erdős: Total Devotion to the Subject." *Science,* April 8, 1977.

Kolata, Gina Bari. "Paul Erdos, 83, a Wayfarer in Math's Vanguard, Is Dead." *New York Times,* September 24, 1996, p. 1.

Krauthammer, C. "Paul Erdős, Sweet Genius." *Washington Post*, September 27, 1996.

Lanouette, William (with Bela Silard). *Genius in the Shadows: a Biography of Leo Szilard, the Man Behind the Bomb*. New York: Charles Scribner's Sons, 1992.

Lukacs, John. *Budapest 1900: A Historical Portrait of a City & Its Culture*. New York: Grove Press, 1988.

MacHale, Desmond. *Conic Sections*. Dublin: Boole Press, 1993.

Macrae, Norman. *John von Neumann*. New York: Pantheon, 1992.

Marx, George. *The Voice of the Martians*. Budapest: Akadémiai Kiadó, 1994.

Miklós, D., et al. *Combinatorics: Paul Erdős Is Eighty*. Bolyai Society Mathematical Studies, Vol. 2. Budapest: János Bolyai Mathematical Society, 1996.

Muir, Jane. *Of Men & Numbers: The Story of the Great Mathematicians*. New York: Dover Publications, Inc., 1961.

Nelson, Carl, et al. "714 and 715." *J. Recreational Mathematics*, vol. 7, no. 2 (Spring 1974), pp. 87–89.

P., B. "Should All International Congresses Be Held Abroad?" *Science*, March 19, 1954, p. 3A.

Pach, János. "Two Places at Once: A Remembrance of Paul Erdős." *The Mathematical Intelligencer*, vol. 19, no. 2 (Spring 1997), pp. 38–49.

Paulos, John Allen. *Mathematics and Humor*. Chicago: The University of Chicago Press, 1980.

Pearson, R. "Paul Erdős, an Eccentric Titan of Mathematical Theory, Dies." *Washington Post*, September 24, 1996.

Peterson, I. "Paul Erdős: An Infinity of Problems." *Science News*, October 5, 1996.

Peterson, I. "Progressing to a Set of Consecutive Primes." *Science News*, September 9, 1995, pp. 167ff.

Pier, Jean-Paul, ed. *Development of Mathematics 1900–1950*. Basel: Birkhäuser Verlag, 1994.

Pinker, Steven. *How the Mind Works*. New York: W. W. Norton & Company, 1997.

Radó, R. "Paul Erdős Is Seventy Years Old." *Combinatorica* 3 (3–4, 1983), pp. 243–44.

Ramanujan, Srinivasa. *Collected Papers of Srinivasa Ramanujan.* Ed. G. H. Hardy et al. Cambridge: Cambridge University Press, 1927.

Regis, Ed. *Who Got Einstein's Office? Eccentricity and Genius at the Institute for Advanced Study.* Reading: Addison-Wesley Publishing Company, 1987.

Reid, Constance. *Hilbert.* New York: Copernicus, 1970.

Rhodes, Richard. *The Making of the Atomic Bomb.* New York: Simon and Schuster, 1986.

Ribenboim, Paulo. *The New Book of Prime Number Records.* New York: Springer, 1995.

Rota, Gian-Carlo. *Indiscrete Thoughts.* Boston: Birkhäuser, 1997.

Schechter, Bruce. "A Prime Discovery." *Discover,* January 1981, pp. 30–31.

Schechter, Bruce. "Ronald L. Graham: The Peripatetic Number Juggler." *Discover,* October 1982, pp. 45–52.

Schmalz, Rosemary. *Out of the Mouths of Mathematicians: A Quotation Book for Philomaths.* Washington, D.C.: MAA Spectrum, 1993.

Singh, Simo. *Fermat's Enigma: The Epic Quest to Solve the World's Greatest Math Problem.* New York: Walker and Company, 1997.

Steen, Lynn Arthur, ed. *Mathematics Today: Twelve Informal Essays.* New York: Springer-Verlag, 1978.

Straus, E. G. "Paul Erdős at 70." *Combinatorica* 3 (3–4, 1983), pp. 245–246.

Struve, Otto. "Scientists and the McCarran Act." *Science,* September 17, 1954, pp. 465–466.

Sved, Marta. "Paul Erdős—Portrait of Our New Academician." *Journal of the Australian Mathematical Society.* Adelaide: University of Adelaide, 1987.

Szénássy, Barna. *History of Mathematics in Hungary until the 20th Century.* New York: Springer-Verlag, 1992.

Tierney, John. "Paul Erdős Is in Town. His Brain Is Open." *Science* 84, October, pp. 40–47.

Turán, Pál. "The Life of Alfréd Rényi." *Matematkai Iapok. Bolyai János Matematikai tórsulat,* vol. 21 (1970), pp. 199–210.

Ulam, Stanislaw M. *Adventures of a Mathematician.* Berkeley: University of California Press, 1976.

van Lint, J. H., and R. M. Wilson. *A Course in Combinatorics.* Cambridge: Cambridge University Press, 1992.

Whitehead, C., ed. *Surveys in Combinatorics, 1987: Invited Papers for the Eleventh British Combinatorial Conference.* (11th British Combinatorial Conference, Goldsmiths' College, 1987.) Cambridge: Cambridge University Press, 1987.

Einstein, Albert, 25, 40, 102, 104, 116, 117, 130, 140, 145, 178–79, 184
Elementary proofs, 134, 143
Encke, Johann Franz, 139
Engländer, Lajos. *See* Erdős, Lajos
Enigma Machine, 96
Eötvös, József, 25
Eötvös, Roland, 25
Epsilons (children), 69, 70, 79, 129–30, 171, 172, 174, 189–90, 195
Eratosthenes, 134–36
Erdős, meaning of name, 24
Erdős, Anna Wilhelm, 24, 25–26, 42–43, 46, 66–67, 118, 131–32, 160, 162, 164, 169, 170, 172, 187–90
Erdős, Klára, 25–26
Erdős, Lajos, 23–26, 41–42, 43–46, 119
Erdős, Magda, 25–26
Erdős, Paul. *See also* specific mathematical concepts and proofs
    amphetamines used by, 189–90, 196
    arrival of, stating, "My brain is open," 13–14, 203
    atheism of, 70–71, 161
    awareness of own mortality, 40–41, 196, 201–202
    on being Jewish, 42–43, 58
    birth of, 26
    cash prizes offered by, 19, 65n, 193–194
    as child prodigy, 15, 20–21, 26, 27–30, 34, 48, 51
    and collaboration, 14, 55, 60, 149, 178–79, 181–83
    correspondence of, 177, 190
    death of and memorial service for, 201–203
    deaths of sisters of, 26
    devotion to mathematical pursuit of pure beauty, 186–87
    dizziness experienced by, 169, 192
    education of, 44, 45–46, 54, 56–58
    English spoken by, 46
    exile from U.S., 164–67
    and father's death, 119, 160
    favorite pastimes and interests of, 16–17, 54, 103, 192, 199
    financial affairs of, 17, 118, 130, 190, 193–97
    friendships important to, 68
    generosity of, 17, 165, 186, 202
    government authorities' reactions to, 130–31, 163–65
    health problems of, 197–200
    helplessness in domestic matters, 43, 91, 195
    honors and prizes for, 17, 150, 180–181
    hyperactivity of, 44, 116, 120
    idiosyncrasies of, 65–66, 69–70, 79, 116, 120, 202
    initials appended to name, 41, 189, 196, 196n
    isolation from other people, 68–69
    Israeli citizenship for, 165
    and *KöMal* journal, 54–56, 184
    labeled fellow traveler by U.S., 164–166
    lecturing style of, 59, 86, 196
    and McCarthyism, 162–66
    memory of, 17, 177–78
    and mother's death, 188–89
    notebooks and journals of, 170
    obstinacy of, 42–43, 68, 165, 200
    and parents, 23–26, 27, 41–46, 66–67, 69, 119, 120, 131–32, 160, 162, 164, 169, 170, 172, 187–90
    parents' protectiveness of, 43, 66–67
    PhD received by, 91
    physical appearance of, 13, 56, 120, 192
    physical contact as uncomfortable for, 68–69
    poetry by, 16, 183, 201
    at Princeton Institute for Advanced Study, 99–104, 111–18, 119
    protégés of, 15–16, 55, 168–75, 185–186
    relationships with women, 67–68, 79
    sexuality of, 67–69, 130
    short university appointments and lectureships of, 130, 144, 161
    travels of, 13–14, 91, 99, 151–53, 165–167, 176–77, 187–88, 197, 199, 201
    trips to Hungary after World War II, 160–61, 167–75, 190
    university education of, 56–58, 91–92

Hevesy, George de, 22, 25
"Highly irregular" graphs, 197
Hilbert, David, 81, 126–27, 142, 203
Hilbert space, 103–104
History of mathematics, 30–35, 49, 50–51, 71–73, 79–88, 105–106, 121–122, 134–36, 154
Hitler, Adolf, 91, 100, 118
Hobbes, Thomas, 50
Hoffman, Paul, 147
Holocaust, 131, 132, 153, 160
Houdini, Harry, 23
Houtermans, Fritz, 23
Hua, Lo Ken, 164, 176
Humphrey, Hubert, 166
Hungarian Academy of Sciences, 15, 41, 164, 167, 168, 170, 190, 201
Hungarian Institute of Mathematics, 161
Hungary. *See also* Budapest
  anti-Semitism in, 22, 42, 56, 90–91
  Ausgleich and, 21–22
  brilliant expatriates from, 22–23
  Commune in, 42, 43, 56
  Communist control of, 42
  Danube River in, 21
  economic and intellectual flowering of, 21–22
  educational system in, 23, 24–25
  in Erdős's youth, 19–20
  Erdős's trips to, after World War II, 160–61, 167–75, 190
  German occupation of, in World War II, 131
  invasion of, after World War I, 42
  Jews in, 22, 23–26, 42–43, 56, 58, 90–91, 131
  Numerus Clausus limiting Jewish admissions to universities in, 56
  political situation after World War II, 160–61
  pride in Erdős, 167–68
  and Trianon peace treaty, 56
  during and after World War I, 41–42
Hurewicz, Witold, 103
Hutchins, Robert Maynard, 190
Hydrogen bomb, 20, 22

Imaginary numbers, 141–42
Immigration and Naturalization Service (INS), 163–65, 167
Inaccessible cardinals, 127
India, 15–16, 17, 60–61, 128, 187–88
Infants and arithmetic, 28
Infeld, Leopold, 116
Infinite sets, 121, 122–25, 127
Infinitude of primes, 51–52
Infinity theory, 73
Integer sequences, 97
Integers, product of, 104
International Congress of Mathematicians, 163, 165
International Congress of Psychology, 163
Internet, 49, 157, 182
Interpolation theory, 116, 133
Irrational numbers, 34, 37, 39
Israel, 17, 165, 199

Jargon. *See* Erdősese
Jews
  and Anti-Semitism, 22, 42, 56, 90–91, 153
  conversion of, 42–43
  Erdős on being Jewish, 42–43, 58
  Erdős's mother as, 24, 42–43
  and Holocaust, 131, 132, 153, 160
  in Hungary, 22, 23–26, 42–43, 56, 58, 90–91, 131
  "neologs," 23
  and *Numerus Clausus*, 56, 153
Johns Hopkins, 104, 117
Johnson, Paul, 96
Jordan's theorem, 69

Kac, Mark, 104–105, 110–12, 133
Kaczynski, Theodore, 18
Kakutani, Shizuo, 130, 131
Kálmár, László, 60, 61–62, 168
Kanigel, Robert, 61, 137, 180
Kármán, Mór, 25
Kármán, Theodor von, 22, 24, 25
Károly, Count·Mihály, 42
Kasparov, Gary, 103
Kauffman, Stuart, 159–60
Kemény, Sándor, 64
Keynes, John Maynard, 80–81

Prime numbers (*cont.*)
  Felkel's work on, 136
  first 100 primes, 48
  Gauss's work on, 136–40
  GIMPS project on, 49
  Kulik's work on, 136
  largest prime number, 48–49
  Legendre's work on, 139
  Riemann's work on, 140–41
  and Sieve of Eratosthenes, 134–136
  table of, 134–35
Princeton Institute for Advanced study, 99–104, 111–18, 119, 131, 143–44
Prizes offered by Erdős, 19, 65, 193–94
Probabilistic method (Erdős method), 65n, 108, 157
Probabilistic number theory, 112
Probability theory, 81, 104–12, 133, 153, 155, 156–57
Prochus Diadochus, 35
Prodigies. *See* Mathematical prodigies
Proofs
  of Bertrand's Postulate, 58–59
  of Chebychev's theorem, 59–62
  and computer, 82
  of Dirichlet Theorem, 144, 145
  of distance between prime numbers, 52–53
  elegance of, 35
  elementary proofs, 134, 143
  Erdős on best proofs, 35
  Erdős on God's book of, 35, 70
  existence proofs, 40, 65, 65n
  of Fermat's Last Theorem, 35–36, 96
  of infinitude of primes, 51–52
  of Klein's puzzle on convex polygons, 79–80, 88–89
  length of, 36
  in mathematics generally, 14, 18
  of Prime Number Theorem, 134, 143–51
  Pythagoras on, 33–34
  of Pythagorean theorem, 20–21, 33–34
  Rota on, 35

of Shur's conjecture on abundant numbers, 63
of square root of two, 34, 36–40
and Thales, 34
Protégés of Erdős, 15–16, 55, 168–75, 185–86
Purdue University, 130, 166, 181
Pythagoras, 31–40, 46, 51, 63, 149
Pythagorean theorem, 20–21, 32–34
Pythagorean triples, 32–33

Quantum theory, 23, 104, 114

Radioactive tracers, 22
Rado, Richard, 62, 91–92, 97, 99, 127, 178
Rajk, László, 161
Ramanujan, Srinivasa, 17, 60–62, 137, 179–80
Ramsey, Arthur S., 79
Ramsey, Frank Plumpton, 80–88
Ramsey numbers, 156
Ramsey theory, 81–89, 92, 127, 157–158, 173
Rand Corporation, 184–85
Random graphs, 156–60
Rational numbers, 123
Rátz, László, 25, 54
Ravel, Maurice, 199
Real numbers, 123–25
Reductio ad absurdum, 24, 40, 123
Regis, Ed, 115, 116
Regius, Hudalricus, 49
Regular graphs, 197
Reid, Constance, 142–43
Reiner, Fritz, 23
Relativity theory, 25
Rényi, Alfréd, 152–56, 158, 161, 168, 170, 171, 184
Rényi, Catherine, 170
Rhodes, Richard, 22
Riemann, Bernhard, 18, 140–41
Riemann integrals, 140
Riemann zeta function, 140–43
Rips, Eliyahu, 86
Roentgen, Wilhelm, 99
Roots of polynomials, 144
Ross, Arnold, 161, 162, 166
Rota, Gian-Carlo, 19, 35, 59